BLACK
SUN

www.penguin.co.uk

BLACK
SUN

OWEN
MATTHEWS

BANTAM PRESS

TRANSWORLD PUBLISHERS
61–63 Uxbridge Road, London W5 5SA
www.penguin.co.uk

Transworld is part of the Penguin Random House group of companies
whose addresses can be found at global.penguinrandomhouse.com

First published in Great Britain in 2019 by Bantam Press
an imprint of Transworld Publishers

A CIP catalogue record for this book
is available from the British Library.

ISBNs
9781787631823 (hb)
9781787631830 (tpb)

Typeset in 12/15 pt Adobe Garamond Pro by Jouve (UK), Milton Keynes
Printed and bound in Great Britain by Clays Ltd, Elcograf S.p.A.

Penguin Random House is committed to a sustainable future
for our business, our readers and our planet. This book is made
from Forest Stewardship Council® certified paper.

3 5 7 9 10 8 6 4

To Xenia,
Nikita, and
Teddy

CONTENTS

PROLOGUE

The air-raid siren sounded at dawn. Its rising wail was relayed across the sleeping town by loudspeakers mounted on lampposts, in the corridors of dormitories and barracks, and in the entrance halls of laboratories and workshops. It reverberated from the abandoned church belfry that faced Lenin Square, sending flights of startled pigeons up into the gray October morning. The birds wheeled over the rooftops of the old town center, over the new parks and apartment buildings, over guard towers and the three concentric rings of barbed wire. Finally they flapped over the dark forest that encircled the secret city of Arzamas-16 like a sea.

In the main machine hall, the whining lathes slowed to a whir. Banks of fluorescent lights snapped off, leaving the operators blinking in morning light that filtered through the glass roof. In the parachute workshop, needles nodded to a halt between the seamstresses' spread fingers. The women straightened stiffly, grateful for the weekly air-raid drill and an early end to their night shift. In the blueprint room, tousled young engineers swept Lucite rulers and set-squares off their drawing tables, rolled plans into long asbestos tubes, and clattered down the stairs toward a row of fireproof safes.

Fifty meters below their feet, a squad of soldiers ran, crooked with sleep, to their battle stations outside the main warhead vault. White-coated men filed out of the bunker chatting, patting pockets for matches and cigarettes. Behind them they left orderly rows of lead canisters stacked in cubicles, a large steel hemisphere sprouting wires, vessels of dull metal as big as bathtubs. Once the last of the scientists had exited, the soldiers hauled the steel blast door shut behind them. Their commanding officer rolled the bolts home with a soft clang.

Alone in its secret vault, deep in the bowels of the All-Union Scientific Research Institute of Experimental Physics, the bomb they called RDS-220 stood alone in silence and darkness.

On his blood-soaked sheets, Fyodor Petrov did not stir. He heard the siren's wail as a rising swell on the furthest edge of his consciousness. All night he had been rafting across a sea of pain, rolled by nausea. Liquid fire was consuming his body.

Now, Petrov saw light. He remembered that light has mass, and exerts pressure. A physical pressure, tiny but measurable. He seemed to feel its particles as they fell on the skin of his face, streaming toward him from the surface of the sun. He tried to rise against the light, but his young body would not obey him. He willed one hand into motion. It jerked spastically as it crawled up his torso. His face was stuck to his pillow. His fingers scraped at a tacky, fibrous mass under his cheek and raised a pinch to his unfocused eyes. His own blond hair, shed in the night, matted with blood and vomit.

'But I can't die,' Petrov heard his own voice argue. 'If I die, I will never know.'

Petrov let his hand drop. Numb darkness spread over him.

He dreamt of fire, consuming the world in a furious tornado. He saw the proud towers of the Kremlin torn from their foundations, disintegrating into ziggurats of dust. He saw boiling seas and bending forests exploding into flame. The whole earth burning, at his command.

The faces of his teachers, friends, and comrades rose before him. They were arguing among themselves, but he could not understand what they were saying. Lost deep inside himself, Petrov felt the

outside world dissolve. The flesh that had clung to him so tortu-
ously all night finally fell away. He had become a spirit, rising
vertiginously into space with a cold wind rushing on his face. Delivered at last into infinite peace, a billion stars inside his head blazed
into light.

The siren stopped. And with it, so did Fyodor Petrov's weak
human heart.

PART ONE

THE CITY THAT DOESN'T EXIST

What do we mean by 'understanding' something? We can imagine that this complicated array of moving things which constitutes 'the world' is something like a great chess game being played by the gods, and we are observers of the game. We do not know what the rules of the game are; all we are allowed to do is to watch the playing.

RICHARD FEYNMAN

CHAPTER ONE

SATURDAY, 21 OCTOBER 1961

NINE DAYS BEFORE THE TEST

I

The train jerked to a halt, jolting Alexander Vasin out of his crumpled doze. In the opposite corner of the compartment, the dough-faced Party man who had traveled with him from Moscow without a word snored softly on, arms folded across his chest.

Outside, the autumn night was still and moonless. The train had stopped in a no-man's-land enclosed by two long walls of barbed wire, illuminated by rows of electric lights. A strip of freshly raked sand stretched into the darkness. Somewhere up ahead Vasin could hear the barking of guard dogs.

He breathed in the fragrant silence. This train was like no other he had ever traveled on. The compartment was brand-new soft-class rolling stock. It was redolent of the future: leatherette and Formica and rubber sealant. An automatic ventilator blew warm air gently onto his ankles. Vasin stepped gingerly over the apparatchik's outstretched legs and pulled open the sliding door.

The trains of his childhood had been like mobile villages, full of chatter, crying, arguments. Lurching theaters of humanity, cluttered with suitcases and leaking bedrolls. But this one was silent, smooth-running, and as hermetic as a spaceship. Only at the vestibule at the end of the carriage did the chilly night air reach in, bringing the familiar train smell of coal smoke and damp grass. Vasin shivered and buttoned his prickly new uniform tunic, retrieving a packet of Orbita cigarettes from the pocket. Orbita: fashionable,

7

hard to find, strong. An apparatchik's cigarette. Better than he'd been used to.

Vasin straightened his uniform in the glass of the door. He had his father's high forehead, dark blond hair just starting to recede. He tucked his new spectacles into his top pocket and squinted again, smoothing his hair and flexing his shoulders to fill out the tunic. Bars of rank on his collar, a sword-and-shield emblem on his right breast. Major Vasin, KGB.

From the corridor came a low murmur of voices in another compartment. A muffled dance tune began, midsong, from a radio in the conductress's cubbyhole. There was a shush of escaping steam and the screech of spinning wheels as the train resumed its motion. It trundled through a floodlit checkpoint into a long barbed-wire cage supported by a timber frame. A pair of barking Alsatians choked on their leads as they stood on their back legs, almost pulling their handlers off their feet.

In the distance the lights of a city appeared, the hard urban crenellations of tower blocks. A single-platform station slid into view.

Vasin hurried back to the compartment, disturbing his companion in the middle of a mighty yawn. He waited in the doorway for the older man to pull on a thick mackintosh and slip a plastic suitcase from the shelf. He gave a curt parting nod as the train slowed to a halt.

Up and down the carriage, compartment doors were sliding open. Vasin wrestled down his large prewar Bakelite case, a prized family possession. He waited for his fellow passengers to pass before he hauled it onto the platform. The young conductress stood by the door smiling, pert and pretty in her uniform coat, her fore-and-aft cap perched on a pile of peroxided hair.

At the stationmaster's whistle the locomotive reversed away from the platform, the red star emblazoned on the front of its boiler disappearing into the night. The new arrivals were momentarily blanketed in a cloud of hot, oil-scented steam.

The guards saluted every passenger and requested papers, corralling them to a pair of clerks, who sat checking and stamping in a bright pool of lamplight. To Vasin's surprise they made no attempt to search any luggage.

In the empty waiting room a stocky, bearded man sat hunched on a bench, holding a book close to his face. He wore a creased trilby hat, and his winter boots were half-laced and unpolished. Vasin stood before him in slightly bemused silence.

'Ah! Comrade Major Vasin?' The man stood quickly, snapping the book shut and scooping it into his coat pocket. 'Greetings. Vadim Kuznetsov. Major. Arzamas State Security.'

He was a head shorter than Vasin, but nonetheless contrived to look down his long nose at him, squinting through black-rimmed glasses. His shirt was buttoned tightly round a thick neck, and his pointed beard jutted forward.

'Welcome to Arzamas-16. The city that does not exist.'

Outside the station, the last of Vasin's fellow passengers were boarding a small bus. The only other vehicle standing on the forecourt was a UAZ military jeep.

'This is us.'

Kuznetsov jerked down the stiff door handle and tossed Vasin's suitcase unceremoniously onto the backseat.

'Jump in.'

'You don't lock the car?'

'Ha! No thieves in Arzamas! This is the most honest city in the Soviet Union.'

Kuznetsov bounced into the driver's seat, pumped the accelerator, and held up a finger, demanding reverential silence. The engine shuddered into life.

'Miracles!'

He ground the jeep into first gear.

'She's not broken in yet. You know, new cars.'

Vasin glanced sharply at his companion for any sign of mockery. But Kuznetsov was oblivious, wrestling the UAZ's gear stick. His was evidently a world where new cars were an everyday annoyance. They accelerated alarmingly along a broad, freshly tarmaced boulevard.

'Our beautiful town.' Kuznetsov waved a hand airily as he zoomed through a crossroads without slowing down or looking for crossing traffic. 'We'll have the scenic tour tomorrow.'

They saw no other people or cars as the city thinned from stucco pre-Revolutionary facades around the station into uniform rows of the modern five-story concrete blocks known as Khrushchevki.

'Here we are. You'll be staying with me for your visit.'

The engine shuddered to a halt. The night was still except for the croaking of frogs. A row of young apple trees gave off a strong odor of rotting fruit.

Kuznetsov's apartment was large and empty. A broad corridor ended in a deep bookshelf, on which a few books were haphazardly stacked. On the right were two spacious rooms; on the left was a sitting room and, beyond it, a kitchen and bathroom.

'I'm in here. You're next door.'

Kuznetsov gestured casually into the first of the bedrooms, where a mulch of shirts and coat hangers covered the bed and spilled onto the floor and a small desk stood covered in notes and printed papers.

In the sitting room, polished glass-fronted cupboards entirely filled one wall. It looked like the House of the Future exhibition Vasin had visited with his son, Nikita, at the start of the summer holiday: boxy armchairs and a square sofa, upholstered in bright-striped fabric. Not a sofa bed, but a compact two-person sofa which could not be used for sleeping on. Vasin had never seen such a thing. Before his marriage, he and his mother had lived in two adjacent rooms in a rambling, high-ceilinged communal apartment off Metrostroyevskaya Street in Moscow. They shared the kitchen and bathroom with two other families, seven people in all. With his transfer to the KGB, Vasin had moved with Vera and their son into a two-room apartment of their own near Gorky Park, a sign of giddying privilege. Yet here he stood in a room in which no one lived at all. A room just for sitting in. In the corner stood a large radio and record player, the latest model from Rigonda, in an oak case. And on the shelf a meter-long row of records.

'From Czechoslovakia,' Kuznetsov called from the kitchen. 'The furniture, I mean. They brought a trainload of it last year. Nice, no?'

Vasin's agreement was drowned out by a clatter of pans.

'Got some food from the canteen. Borscht. Meatballs. Mashed potatoes.'

In the kitchen a new refrigerator, not a rumbling monster from the Stalin Factory, purred in the corner. Kuznetsov tossed Army-style aluminum mess tins onto the Formica kitchen table.

'We'll have everything, I guess? I'm hungry too.'

Kuznetsov set enameled pots onto the electric cooker with a clatter and unscrewed the mess tin lids.

'Go wash if you like. I'm an excellent cook. Look!'

He took a mess tin in each hand and splashed their contents into the pots.

By the time Vasin returned from his shower Kuznetsov was hunched over the table, slurping soup. A portion for Vasin steamed in a large Uzbek bowl.

'So. Has someone briefed you on what happened?' Kuznetsov pointed his beard quizzically at his new roommate.

'The case summary says that Fyodor Petrov was poisoned. Accidentally.'

'Right. Bright young physicist. Sad business.'

'It says a lot about what happened. Nothing about why.'

Kuznetsov pushed away his empty bowl, stood to open the window a crack, and lit a cigarette.

'Why, indeed. That is the question, Comrade Vasin.' He spun a steel ashtray onto the table. 'We are not very used to outsiders here in Arzamas. To what do we owe the pleasure?'

Vasin slowly dipped his spoon in the soup, tasting it in silence. Then: 'Good soup.'

'Wouldn't want you to starve, Comrade.'

Vasin ate on in silence.

'Is there a reason I shouldn't have come, Comrade Kuznetsov?'

'Forgive me. Personnel tells us that you have recently joined State Security—'

'From the Moscow Police Criminal Investigation Department. Homicide Department. That is correct.'

'Homicide?'

'Does that bother you, Comrade?'

'Well. You know about Arzamas. Something about the word "homicide" makes us nervous. And the Lubyanka usually lets us take care of our own business.'

'Actually, I don't know about Arzamas.'

'They didn't tell you anything in Moscow?'

'Let's say they didn't.'

Kuznetsov exhaled smoke.

'Consider Arzamas a separate planet. Some of the greatest minds of the Soviet Union are here, doing vital work for the defense of the Motherland. Social deviants to a man, in the opinion of some of our colleagues.' Kuznetsov leaned forward, his breath strong on Vasin's face. 'But here's the thing – no one cares about what they do. What they think. What they read. Who they sleep with. Nothing matters, as long as they do the job they're here to do. So take all the rules you know, and add a new rule at the top: Nothing interferes with the project.'

'What project?'

Kuznetsov snorted and loudly clattered his bowl into the sink.

'Kuznetsov.' Vasin softened his tone. 'Really. What project?'

'Nothing interferes with RDS-220.'

Vasin digested this for a moment.

'Which is a kind of bomb?'

Kuznetsov flinched at the word.

'It is a device. A new device.'

'Isn't that what they do here? Why this is a secret city? Make new *devices*?'

'Not like this one. It is bigger. Much bigger. And urgent. Top-level Politburo order.'

'Our Soviet way. Always the biggest. Always the best.'

Kuznetsov's mouth contracted into a thin line. His nostrils flared as though sniffing for a hint of mockery. Vasin sensed that he had bumped into a hard edge in his hitherto amiable companion. Kuznetsov waited a long moment before replying, his eyes traveling over Vasin's face.

'We'll find out soon enough.'

'Find out, how?'

'They're going to test it. And before you ask, by *test* I mean detonate the thing.'

'Here?'

'Not *here*, numbnuts.' The tension in Kuznetsov's mouth eased at the obtuseness of Vasin's question. 'They test the devices up in the Arctic. Don't they tell you anything in Moscow?'

'Well, that's a relief.'

'For some.'

'Meaning?'

'Meaning – let's hope so. The farther away that thing is from *me*, the better.'

Kuznetsov fumbled another cigarette into his mouth and testily struck a match.

'And the deceased . . . ?'

'Fyodor Petrov was a key member of the RDS-220 team.'

'Ah.'

'Yes. *Ah*. An assistant to Professor Adamov, no less.'

'The Director?'

'The Director. The Tsar and God of Arzamas. Father of RDS-220.'

'So what do you think happened to Petrov, exactly?'

'You read the file.'

'I want to hear your version.'

Kuznetsov exhaled smoke through his nostrils like a cartoon demon.

'I don't have a *version*, Vasin. The boss, Major General Zaitsev, is going to lay it out for you chapter and verse. A words-of-one-syllable man, our Zaitsev. Tomorrow, 0900 at the *kontora*.'

Kontora. Literally, the office, and one of the more respectful slang words for the KGB. Vasin had heard plenty of others.

13

'Very good. And what have *you* been asked to lay out for me, Kuznetsov?'

'Oh, you know. Bed linen. Towels.'

Kuznetsov cracked a smile, holding Vasin's eye.

'You're funny.'

'So they keep telling me.'

'Seriously, though.'

'Seriously? That's a good word. Shit here *is* serious, Vasin. Which is why I find myself thinking that it may not be my most natural habitat. What I need to lay out for you is this: We do not wish to see Fyodor Petrov's colleagues unduly distracted.'

'Because nothing interferes with the project. I understand. Thanks for filling me in.'

'Pleasure. That's what I'm for. To fill you in.'

'And Petrov's body?'

'Spot the ex-detective. Central Clinical Hospital, I assume. Ask the General tomorrow.'

'Address of the deceased's apartment?'

'No idea. Zaitsev's in charge.'

Probably a lie. But Vasin smiled nonetheless. The address was somewhere in the summary report he had brought from Moscow.

'I'm keeping you up, Kuznetsov. It must be getting late.'

'For Moscow, maybe. Not for the busy bees of Arzamas. We're going out.'

'Out?'

'To a lecture at the All-Union Scientific Research Institute of Experimental Physics. Also known as the Citadel.'

Vasin glanced at his watch.

'A lecture? At eleven at night?'

'Science never sleeps, Comrade. Professor Adamov has something to tell the assembled brains of Arzamas. Which probably does not include us. Sorry. I mean me, at least. But we're going anyway. Come, or we'll be late.'

II

It took them less than five minutes to drive through the empty streets of Arzamas and swing into the broad expanse of Kurchatov Square. A freezing mist was rising. The main building of the Citadel loomed like an ocean liner in the thickening fog, its illuminated windows piercing the night. A colonnade of bare concrete pillars supported a jutting roof. It reminded Vasin of a mainline railway terminus.

A row of turnstiles divided the high-ceilinged lobby in half, a more solid version of the entrance of a Moscow metro station. Kuznetsov flashed his red KGB identity card to the sergeant on duty, and Vasin followed suit.

The lecture theater was crowded to overflowing. Whispering apologies, Kuznetsov pushed his way into the darkness. A couple of young uniformed men shifted up to allow him and Vasin to sit on the carpeted stair. A single table lamp on the raised stage provided the only light, illuminating a lectern. Professor Adamov's gaunt face, lit from below, looked to Vasin like a speaking skull. He wore an old-fashioned black Party member's tunic, buttoned to the collar, with three Hero of the Soviet Union stars pinned over his heart.

'. . . First and foremost, of course, we must be careful. I promised the General Secretary that we would not . . .' Adamov paused, wheezing a little, as he worked up to the punch line. 'That we would not crack the earth like an egg!'

The Professor looked up from his notes and squinted through small glasses at the attentive young faces arranged around him like disciples. He stretched his thin mouth into something like a smile, authorizing a ripple of awkward laughter to spread around the hundred men in the overheated hall.

The Professor's smile widened, even as his mind seemed to turn inward. Silence deepened. It seemed to Vasin as though Adamov was suddenly too busy thinking to acknowledge the outside world. Every other function of his body except breathing appeared suspended while the brain raced at full speed. Vasin glanced left and right, but Adamov's pause seemed to excite no surprise in his audience. Was the quiet filled with racing calculations? Or was Adamov's silence within himself – just silence?

And then, just as abruptly as it had left, animation returned to Adamov's eyes. He peered into the hall at his students and colleagues and examined them, one by one. The faces were eager, open, the eyes shining with intelligence. A few tried to hold his stare, at least for a moment. Some looked brave, some hopeful, most fearful. And then they looked down.

'Every one of you has been chosen.' Adamov's voice was so quiet as to be almost inaudible, as though he were speaking half to himself. 'Chosen for some aspect of your minds that the Motherland has found useful. Or interesting. Or just uselessly unusual. In physics one always has to keep the useless results in mind.'

Another dry joke? If so, this time nobody laughed. Adamov stepped toward a large box that stood at the front of the stage and flicked a switch, illuminating a white square of light on a large screen. An overhead projector, the first Vasin had ever seen.

'Every day newcomers arrive to assist in the final assembly of RDS-220. To them, welcome. And I have an announcement to make to all of you. After receiving the reports of all the laboratory heads, I have concluded that all is in place to finally set a date for the test that we have all been anticipating so eagerly. The date is October thirtieth. We have reached the final stage of preparation. Nine days from now, the world will see the might, the glory, and the genius of peace-loving Soviet science.'

A low murmur ran through the hall. Adamov stabbed the whispering down with an icy glance.

'As I was saying. For the newcomers, and for all of us, a reminder of our fundamental questions. This device has some . . . new features. The consequences of this test will be hard to predict.'

He began writing formulas with a scarlet marker on a transparent sheet on the projector's screen, tapping the point on the glass for emphasis. 'It. Is. Our. Patriotic. Task. To. Calculate. Them . . .

There are great unknowns we have yet to grasp. Consider the work of Dr Smirnov on fusing hydrogen nuclei with their heavy brothers tritium and deuterium. The behavior of superheated plasma, gases hotter than the heart of the sun, during the milliseconds after core detonation. The consequences of scaling up the tried-and-tested thermonuclear reactions to a hitherto unknown scale. What happens when we double it? Multiply it by ten? A thousand? At this kind of scale, gentlemen, we encounter new parameters: the solidity of the earth's crust. The behavior of the atmosphere in different thermoclines. The point at which we may ignite a chain reaction in atmospheric water. And this is the point that we address today. But first, for the benefit of the newcomers, we remind ourselves of some ancient history. Our old friend RDS-100. Back in 1951.'

There was a squeak as the lectern light was dimmed. A film projector clattered into action. A bright white rectangle flung a series of numbers onto the screen, counting down.

'So, colleagues.' When Adamov spoke loudly the pitch of his voice also rose. 'We will remind ourselves of the terrible forces that we believe are under our command. We watch. And we are humbled. You will see the film at one-twenty-fourth speed, frame by frame, so that we can visually establish the detonation stages of this device from ten years ago.'

A rumble began in Adamov's chest that sounded like the start of a phlegmy smoker's cough. The Professor rooted in the pockets of his tunic, drew out a crumpled packet of cigarettes, and struck a match.

The view from the hatch of an aircraft flickered onto the screen. In one corner, a tail fin intruded on the shot. Below, a landscape of whiteness. Sea ice, the outline of a sweeping bay with indistinct shapes on the horizon. An ungainly black shape tumbled earthward, neatly deploying a parachute a couple of seconds into its descent, then gaining stability as it drifted gently down. For nearly a minute, there was nothing but the projector's whir. Then, a sudden flash, making the screen an almost perfect blank for several seconds.

'Now. Slow, please.'

On the screen, as the flash died, smoke rippled centrifugally. A small vertical blast of debris, as from a conventional shell, burst upward. Then a second horizontal ripple, and a third, each raising a ridge of earth and snow as it hurtled out across the landscape. A column of smoke rose and thickened, obscuring the detonation point. Light flashed inside the column as it rose. Then came another detonation, inside the cloud this time and far above ground zero, making the rising smoke suddenly bulge. The frames clicked by. The blast wave reached the aircraft and caused it to lurch crazily for several seconds before recovering. The cloud was level with the aircraft now, and climbing, and spreading. The cameraman pulled the focus back to encompass a vast mushroom of debris spread across the sky.

Vasin found no words for what he was watching. He turned to Kuznetsov, but his companion's attention was still rooted to the now-static final image on the screen, transfixed as a child's at a scary movie.

The lights in the hall at the lecture's conclusion robbed Vasin of the anonymity of darkness. Plenty of the men were in uniform, mostly with the crossed-hammers insignia of military engineers. But Vasin's KGB sword-and-shield badges immediately marked him out as an intruder.

The crowd on the stairs shuffled aside to allow Adamov to pass. Vasin felt the Professor's eye catch on the telltale uniform, the officer's bars on his collar, his face. The old man's pale face momentarily creased with distaste.

The Professor moved on up the stairs. Vasin slipped into his wake. He heard Kuznetsov call something after him, and ignored it. Pushing forward among the bodies crushing through the doors with a skill learned on the Moscow metro, Vasin squeezed into the corridor and raced after the retreating figure of the Professor and his entourage of assistants.

'Professor Adamov? A moment, please.'

Vasin's raised voice was enough to stop Adamov in his tracks, if only because it was clearly unheard of for anybody to shout the Professor's name in the halls of the Institute. Catching up with Adamov, he felt the full weight of the Professor's outraged glare.

'Major Alexander Vasin. State Security.'

Adamov did not speak, but stood motionless, waiting for one of his acolytes to interpret his silence. A white-coated youngster consulted a clipboard.

'Professor, there was a letter from the *Kommandatura* this morning. Major Vasin is here to investigate Dr Petrov's accident.'

Vasin saluted.

'My apologies for the disturbance, Professor. But I hope you understand. . . .'

Adamov raised a long-fingered hand in front of Vasin's face, as though stopping traffic. The gesture was imperious.

'A terrible tragedy. But I have spoken to one of you already. Major . . . Efremov? There we are. Thank you. Goodbye.'

Adamov turned to go, his palm still raised rudely in Vasin's face.

'Sir?' Vasin flung the word hard enough to stop Adamov in his tracks once more. 'I am afraid that there will have to be more questions. I have been sent from Moscow on the personal orders of General Orlov to conduct an independent assessment of the case.'

Slowly, Adamov turned back.

'General *Orlov*.' Close up, Adamov's face was gaunt as a corpse's. He spoke slowly, and there was menace in his voice. 'Now if only we had as many *hours* as we have *generals*. And what is it that your general needs from me?'

'Thank you, Professor. May I have the honor of speaking to you in private?'

An indecent hiss escaped the Professor's dry lips.

'What will cost me less? Arguing with your generals, or making time to talk to you?'

'Professor, you answer your own questions so succinctly. Talking to me should take no time at all.'

The Professor's mouth clamped tight as a trap. His pale blue eyes filled with fury.

My God, thought Vasin, his eyes connecting for a long moment with Adamov's wrathful stare. This is a man who can hate.

'Perhaps. After the test.'

Adamov turned his back on Vasin and strode onward.

Vasin felt a strong hand gripping his upper arm. Kuznetsov pulled him to the side of the corridor with enough force to make a point. Young scientists and engineers streamed past them, chatting animatedly. Kuznetsov's voice hissed into his ear.

'What the *fuck* was that?'

Vasin pulled his arm free and turned to his host. His handler.

'I wanted to make an appointment. Is there a problem?'

'A fucking *appointment* with Professor Academician Yury Adamov? Yes, there is a problem.'

'Is he not a witness in the Petrov case?'

'Vasin. So you're a big shot from some top-secret cubbyhole of the *kontora*'s top floor. Orders from above. I see. But Adamov . . .'

The crowd spilling out of the lecture theater pushed them apart for a moment before Kuznetsov could continue.

'. . . Adamov *is* Arzamas. The program is *his*. He is . . .'

'Above the law?'

'He's off-limits to you. To everyone.'

'To *you*, Kuznetsov. Maybe he's off-limits to you.'

Vasin saw a red flush of anger boiling up from Kuznetsov's tight collar like a rising storm. But the man forced it down, like a child fighting to control a tantrum. Kuznetsov exhaled deeply, twice, and when he spoke again his voice was impressively calm.

'Vasin. Alexander. Or may I – Sasha? *Sasha*, listen to me. This place is not like other places. It's not like anyplace you've ever been.'

'You don't know the places I've been.'

'Nowhere in our broad, glorious Union is like Arzamas. Different rules.'

'I believe you. But would you be surprised to know I've heard that before?'

Kuznetsov raised his eyes to the heavens in a pantomime of exasperation.

'I give up. You really need to speak to Zaitsev.'

'I didn't think I had a choice in the matter.'

The two men stared at each other. The corridor had finally emptied. The only sound was the distant clatter of a Teletype machine and the fading chatter of the departing crowd.

Vasin broke the tension first.

'That film was . . .'

Kuznetsov threw him a low glance.

'Terrifying? Yes.'

'You've seen it before?'

'It's Professor Adamov's favorite. It is why I brought you along.'

'And the bomb he's building now. It's . . .'

'Bigger than that. Hundreds of times bigger. See, I wanted to fill you in on what they do here. They make machines to kill the planet.'

CHAPTER TWO

SUNDAY, 22 OCTOBER 1961

EIGHT DAYS BEFORE THE TEST

I

Early the next morning Vasin turned up his mackintosh collar against the rain and mist that drifted down the broad boulevard outside Kuznetsov's apartment building. The sky was heavy with low, dawdling clouds. The slow weather of deep Russia, where seasons follow each other like a procession of steamrollers, trundling and relentless. Autumn was a dripping season of sweet rot and the sound of running water in hidden places.

Kuznetsov's jeep shuddered reluctantly into life. He gunned the engine to attract Vasin's attention.

'Come on, old man. Big shots are waiting for you.'

Kuznetsov dropped him off in front of Arzamas's KGB headquarters, a stubby modern block screened from the street by a row of fir trees. In the forecourt stood a bust of Felix Dzerzhinsky, founder of the Soviet secret police, his bronze face glistening in the rain.

In the lobby secretaries carrying files clicked on high heels across the marble floor. Even on Sundays and holidays, night and day, the *kontora* worked on. A boy-sentry entered Vasin's name in a ledger with painstaking formality. The place had the same thick smell of floor polish and wet overcoats as his office in Moscow. Somewhere two typewriters clicked in busy disunison. A telephone rang, unanswered.

General Zaitsev's secretary had buttery blond dyed hair and a face that seemed to have been disfigured permanently by constant lying.

'The General has been delayed,' she told him archly. 'Wait.'

'Very good. Please tell the Comrade General that I shall take the opportunity to visit the canteen. Downstairs, I imagine?'

A crack of disapproval creased the secretary's makeup.

'Ah! And the latest issue of *Krokodil*! May I?'

Without waiting for an answer Vasin picked up the Soviet Union's best-loved satirical magazine from a low table. He shrugged off his wet mackintosh and hung it, dripping, on the General's coat stand. Then he went in search of coffee.

The basement cafeteria at the tail end of breakfast was almost deserted. Vasin bought himself a sweet roll and a cup of excellent coffee, Cuban, fresh-ground. He settled at a table and began to leaf through the magazine. The usual nonsense: caricatures of drunken workers, comic poems about nagging mothers-in-law, prose sketches of the charms and absurdities of rural life. From the corner of his eye he saw a tall officer in an immaculately pressed uniform with adjutant's braids enter the dining room. The man peered about, spotted him, then strutted across the room like a clockwork toy.

'Comrade Major Vasin.'

It was not a question. The officer sat down heavily opposite him.

'I am Major Oleg Efremov, General Zaitsev's adjutant.'

The officer's pointed gaze made a slow tour of Vasin's face. He took in the glasses, the soft hands, and Vasin's eyes, steady and insolent as they met his.

'The General is waiting for you. If you please.'

In his tight-fitting tunic General Zaitsev looked like a pre-Revolutionary farmhand buttoned uncomfortably into his Sunday outfit. His neck was wider than his face, and he sat with huge, scarred fists clenched on the table like an ogre ready to eat an intruder who has strayed into his kingdom. One of the university-trained milksops who'd come into the service since Stalin died, for instance. Vasin recognized Zaitsev's type. A State Security officer of the old school, who had earned his stars in blood-spattered execution cellars. A man who'd breathed the smell of fresh death.

'The government inspector has come to check on us.'

Zaitsev spoke with a thick country accent and addressed Vasin in the familiar form, like a wayward child.

'No, sir. I have no reason to believe that your work is anything but of the highest quality.'

'I am told you personally approached Professor Adamov last night. But you had not presented your credentials to the local authorities. To me.'

Vasin nodded slowly. Zaitsev's butcher's face. Those joint-popping hands.

'My apologies, General. My credentials are all here.'

Vasin pulled a sheaf of letters from his tunic pocket and held them out. Zaitsev did not take them.

'Listen to me now. This city is governed by a special regime. There are procedures—'

'General,' Vasin interrupted. 'With all respect, my orders are very clear.'

Zaitsev's face flushed a deeper shade of red.

'My investigators have already reached a conclusion.' The General's voice was emphatic as a blow from a billy club. 'The evidence clearly shows that Fyodor Petrov killed himself. The investigation is over. We are filing the report. You are too late.'

Vasin composed his face into a mask of humility.

'Yes, Comrade General.' Vasin had been through this before. By rank, he was a subordinate. But by the authority he represented he was . . . something else. Something that must be hinted at delicately. At first. 'But I have been ordered by the competent authorities to conduct an independent review of the evidence. And you of course would not wish me to disobey my orders. As you are aware, the deceased's father is personally close to many members of the Politburo.'

Zaitsev gave a porcine grunt.

'Review if you have to. We have assembled definitive evidence. But you are not to approach or harass the principal witnesses. They have already been interviewed to my satisfaction. Is that clear?'

'Definitive evidence, sir?'

'Definitive. Petrov died of thallium poisoning. A radioactive heavy metal. He used thallium in his laboratory. Signed for every milligram taken. But he did not *use* every milligram. The records prove it. A substantial quantity of the thallium is missing. Some two thousand milligrams unaccounted for. Is that definitive enough for you, Major?'

'May I be allowed to see the records, General?'

Zaitsev's scowl turned even more venomous. He turned to his adjutant.

'Efremov? The man from Moscow does not believe me. Bring our transcript of the laboratory files.'

Efremov curled his nose as though at a bad smell and obeyed. While he busied himself opening a large steel safe at the back of Zaitsev's office, the General plucked a sheaf of papers from his in-tray and began to read them, demonstratively ignoring Vasin.

'Comrade General? The report you asked for.'

Zaitsev plucked the gray file from his assistant's delicate hand. The cardboard cover creased in the grip of the General's thick fingers.

'Right. Vasin. Here. Look at it. Every sample of thallium Petrov signed out for the last month. There, on the left, every gram he used in his tests. There, in red, the amount unaccounted for. Took a team of five men three days to comb through all the files to get the information. Began immediately after the postmortem report, finished last night.'

Vasin flicked through the columns of numbers, dates, amounts. They meant nothing to him. As Zaitsev knew.

'May I keep this?'

'You may not. As you see, it is marked "Top Secret."'

'And the transcripts of the witness interviews?'

'They will be filed in the registry, in due course. The case file is being collated now. As per our procedures. When it's finished, you will read it. And agree with it.'

'And the body?'

Zaitsev snorted.

'In a secure morgue.'

'When may I be allowed to see it?'

'Never. Too radioactive. The radiation dissolves tissue like a sugar cube in tea. Or so I'm told.'

'And Petrov's apartment?'

'Same story. Sealed.'

Vasin frowned and looked at the floor.

'So, General, if I have understood correctly, I may not in fact *do* anything? Except compose a telegram to Moscow informing them that I have been prevented from carrying out the Politburo's orders. Good day, Comrades. I imagine Moscow will be in touch.'

Vasin placed his sheaf of credentials on Zaitsev's desk, saluted smartly, and turned on his heel without waiting to be dismissed.

'Wait!'

The boss's voice had sunk to a low growl.

'Major. Just do your job and get out of here. Efremov, you can take our guest to the morgue. He wants a sniff of our Arzamas radiation. Take him now.'

Efremov saluted in turn and stalked out of the room, throwing a glance of contempt at Vasin as he passed. Vasin and Zaitsev remained alone.

'My thanks, Comrade General. I will do my job.'

'You have two days, Vasin. *Two.*'

Or what?

Vasin knew better than to ask.

II

Vasin and Efremov walked down Engels Boulevard without speaking. A fine drizzle shrouded the town in a pall of drifting gray. They emerged into the main square, named for Lenin. One side of the square opened onto a high riverbank. Beyond stood a wooded island topped by the tall belfry and onion domes of a former monastery that no one had got round to demolishing. To their left rose the arrogant modern bulk of the Kino-Teatr Moskva, the facade a sloping expanse of plate glass. Inside the cinema's atrium the chandeliers glowed with dingy light against the morning gloom. The only color in the square came from the windows of the Univermag department store. As they passed Vasin dawdled to examine the goods on display. Czech shoes and German overcoats. A large stack of canned Kamchatka crab. In Moscow, such a cornucopia would draw a crowd. But here, citizens were apparently indifferent to the fantastic luxuries piled high in the shopwindow.

And the people. The way they moved was disconcerting. On this singular planet there were no scrums of grunting housewives, shoving forward toward their objects of desire, a departing tram, a fresh chicken. The people of Arzamas strolled about like extras in a film. They were as well dressed as actors, too, even the manual workers in their striped sailors' undershirts and boiler suits. A model town, full of model citizens.

On a street corner a traffic policeman stood hopefully, waiting for some traffic to direct. None came. Efremov turned in to Kurchatov Street. They passed a restaurant with red velour curtains, a hairdresser's shop with its miasma of violet-scented hair spray, food shops with their standard-issue Soviet signs: MEAT. FISH. An electric tram, the new Polish kind that had only just arrived in Moscow, rumbled past on fresh-laid rails. Arzamas's Central Clinical Hospital stood back from the road, a long gray cube.

At the entrance to the hospital Vasin paused to light an Orbita. Efremov waited, but did not light one of his own. Vasin knew the wisdom of numbing the nostrils, recalling the foul mortuaries that marked the beginnings of most of his cases. A stinking cellar in Tashkent from which some Party bigwig had commandeered the refrigeration unit for his dacha. A charnel house in Rostov on Don where bodies were stacked in promiscuous piles in a grotesque parody of an orgy. But as he and Efremov strode down the stairs to the hospital's basement, Vasin's nostrils were filled only with the clean sting of formaldehyde and disinfectant. A doctor in a crisp laboratory coat stepped backward into the corridor. Catching sight of the two officers, he stopped short.

'Good morning . . . Comrades.'

Their uniforms. Black officer's boots, blue breeches, belt and shoulder straps, the telltale KGB green piping on their caps and epaulets. Back in the days when Vasin used to work in his old dark blue police uniform, crumpled and scruffy, people would roll their eyes. Most Soviet citizens viewed ordinary cops as bunglers, sacks of shit tied with belts. The most common nickname for the police was *musor*, 'garbage'. Ever since his move to the KGB, people shrank at the sight of him. Did he enjoy it? Vasin looked the doctor up and down. A part of him did. The world bends around an officer of State Security. It was like a law of physics, radio waves curving in a magnetic field. It bends – though not usually in the direction of truth.

'Comrade Doctor Andreyev.'

'Major . . . ?'

'Efremov. I have brought one Major Vasin of State Security, from Special Cases in Moscow. He has come to discuss the tragic accident of Fyodor Petrov.'

'Ah.' Dr Andreyev's face eased a little. 'Of course.'

Men still feared the uniform. *Show me the man and I'll show you the crime*, old KGB bruisers of the Stalin generation used to say. Sure, the country had a different leader now and was heading into a different future. Officially the old days of State terror, of indiscriminate arrest lists and regional quotas for executions, had been jettisoned. Or so Vasin chose to believe. Nonetheless the reflex of fear lingered like the ache of an old scar.

'Would you like your visitor to see the pathology report, Major?'

Vasin spoke up.

'And the body.'

Andreyev hesitated.

'Are you aware of the necessary precautions . . . and the risk?'

Vasin nodded grimly. Never admit to ignorance. Andreyev glanced nervously at Efremov, who grimaced his assent.

'Please, go ahead. Our Moscow visitor seems very eager. But if you don't mind I will wait outside.'

'Very well. I will summon my personnel.'

'Is the risk . . . unusually high?'

'Yes, Comrade Major. You will see it in the pathology report. Tests show that young Petrov has enough thallium inside him to poison a city.'

The rough cotton of the oversize overalls chafed Vasin's crotch and made him walk bowlegged. The curved plastic of his face mask was misted with condensation. Andreyev, leading, walked stiffly into a room covered in shining white tiles and illuminated by a powerful surgical lamp. A pair of orderlies, also dressed as spacemen, rolled a dull metal coffin in on a gurney. They struggled to lift off the lid, which came off in weighty sections.

'Lead,' Andreyev called through the rubberized canvas of his mask. 'Lead! Absorbs radiation.'

In the coffin lay a drowned man. Or at least that was Vasin's first impression. The face was bloated, the skin pale and blotched, the eyes and mouth wide open. Petrov's hair had fallen out in clumps and had continued to shed into his coffin. The young man's teeth, too, were loose and covered in clotted blood. Around Petrov's shoulders and chest were scratch marks, as though made by fingernails. Vasin gestured a question with a gloved hand. The doctor mimed tearing off his overalls.

'Self-inflicted. He shredded his clothes.'

The handsome young man in Petrov's file photograph was unrecognizable. In death the victim looked . . . Vasin searched for the word to describe it. Exploded. Petrov's body seemed to have burst like an overboiled sausage.

Unusually, the torso was untouched. Vasin mimed cutting up and sewing together above the stomach. Andreyev wagged a finger.

'No autopsy, Major. Too dangerous,' came his muffled words.

The dead were often Vasin's best informants. Most of his fellow detectives preferred living witnesses that they could browbeat and terrorize. But Vasin knew that dead men most certainly could tell tales. And unlike the living, they rarely lied. Petrov's corpse, however, would keep its secrets locked inside.

'Close it.' Vasin flapped his hands. 'Close it.'

The pathologist eased a black Bakelite device into the space beside the corpse's head – evidently a Geiger counter for measuring radiation. The needles on the dials leapt to maximum and stayed there. Andreyev turned some buttons, coaxing the needles down ward, and took a final reading. Orderlies reappeared, moving quickly, sealing off Petrov's pale blue eyes from the light for the last time.

Vasin and Andreyev filed out through a door different from the one they had entered. Three moon-men awaited them, armed with powerful spray guns. They buffeted Andreyev and Vasin unceremoniously from every direction with hot water, two spraying and the other brushing vigorously with a long-handled broom. Then the white ghosts stripped off Vasin's and Andreyev's protective clothing and pointed them, dripping in their underwear, into a shower room. Even as steam rose around the two men's bodies, Vasin found himself shivering.

'You find our procedures thorough, I hope?'

'I trust this was not only for my benefit, Doctor.'

'We take radiation very seriously at Arzamas.'

'There is no doubt about the cause of death?'

'None. The symptoms are very clear. Petrov ingested a highly radioactive substance sometime last Monday. A simple analysis of his vomit confirmed the presence of thallium. And tissue samples show that he consumed around two thousand milligrams. Two grams. A fatal dose is only around a quarter of one milligram. Therefore he ingested enough to kill eight thousand people. You see why we are reluctant to open him up.'

'And the source of the thallium? Who has access to it?'

Andreyev turned to the investigator.

'Hundreds of people. This whole city is built on radioactive materials. And their uses.'

The doctor tugged up his braces and slipped on his white lab coat.

'Petrov had access?'

'Of course. He worked in the Institute. But you'd have to ask his lab clerks for details. They keep a log, I imagine, in the laboratory.'

'And you, Doctor, what is your feeling about the cause of death?'

'I have no feelings, Comrade Major. Only observations. And my observation is that men who work with reagents such as thallium are professionals. They are well aware of the dangers.'

Vasin now regretted the uniform. Pathologists often had good hunches, usually shared like postcoital endearments over an after-autopsy cigarette. But here in the bright sterility of this hospital basement, there were no dark corners in which confidences could grow.

'Does it look like a suicide to you?'

Andreyev gave Vasin a long look.

'Comrade. The scientists here live in a cloud. But the cloud is small and very high up. And sometimes the cloud gets very crowded. People fall off.'

'Or jump off?'

'That, Comrade, if you will permit me to say, is your department.'

Andreyev shook Vasin's hand and left him standing in the changing room. In a glass window in the laboratory door, Efremov's face appeared, peering in impatiently to see what was keeping Vasin.

III

Outside the hospital Vasin sucked greedily on another cigarette.

'Why didn't you view the body with us, Efremov? You don't seem the squeamish type to me.'

The adjutant, his arms deep in the pockets of his raincoat, merely nodded.

'How long are you going to keep up this strong, silent act, Efremov?'

His companion smiled coldly.

'Are you bored already by Arzamas, Major? In need of conversation?'

'I need information.'

'Such as?'

'Such as, how did Petrov die?'

'It's . . .'

'In the file. Of course. But my memory is terrible. Remind me.'

'Petrov was found dead in his apartment. Killed by thallium poisoning.'

'And what did he do in the last hours of his life?'

'Petrov was last seen alive at dinner with colleagues.'

'Which colleagues?'

'He dined with Professor Adamov and his wife at their home. They reported that Petrov seemed tired but otherwise normal.'

'Was anyone else at dinner?'

'An engineer colonel. Pavel Korin.'

'And how long did it take the thallium to kill Petrov? Any idea when he ingested it? Or how?'

'A matter of hours. He took it himself.'

'You suppose. Did anybody visit him at his apartment after dinner?'

'No.'

'Does his building have a concierge? A guard?'

'He was asleep. It's in his witness statement.'

'So we have no way of knowing if anyone came or left during the night?'

Efremov sighed wearily.

'Petrov took his own life, Vasin. People usually do that alone.'

'Did he leave a note? Can we visit his apartment?'

'Your memory really *is* terrible, Major. General Zaitsev just told you that it was impossible. Too radioactive.'

'He said the same about seeing the body. Yet here we are.'

'You may look at the investigator's photographs.'

'I will. But did you see the apartment yourself?'

Efremov's icy face registered a twitch of emotion.

'I did, as it happens.'

'And what did you see?'

'Blood and radioactive . . .' Efremov seemed to search for a more delicate word but decided against it. 'Radioactive vomit. Everywhere.'

'And where was Petrov?'

Efremov struggled for a moment, torn between distrust and a desire to talk.

'Come on, old man. We're on the same side.'

'Petrov was tangled in his sheets. He'd ripped them into shreds. And he'd torn the pillow apart with his teeth. There was even blood up the wall.'

'Sounds like a pretty horrible way to die.'

Efremov shuddered involuntarily but said nothing for a long moment.

'Maybe he deserved it.'

'*Deserved* it?'

Efremov summoned another glacial smile.

'Right. Enough chitchat.' The adjutant's voice had become brisk and official. He tugged his tunic straight and looked at his watch. 'Registry should be ready for you now. Let's get you buried in that paperwork.'

'Before you bury me . . .'

Efremov's eyes narrowed in suspicion.

'I need to send a telegram, internal and secure. To my boss.'

Secure naturally meaning – to be immediately shown to Zaitsev.

'Telegram?'

'It's time to check in with Moscow. Procedure. My chief likes to keep his finger on the pulse. Unless you'd rather I didn't, of course.'

'Of course.'

Vasin knew that just four words would probably do the trick. REQUEST IMMEDIATE INTERVIEW ADAMOV. If he had learned anything in his year at Special Cases, it was that General Orlov possessed an almost supernatural knack of making some of the most powerful men in the USSR jump to his will. General Zaitsev be damned. Within hours, Vasin guessed, some mighty voice of authority would be on the line instructing the Professor to make time. Now.

IV

Petrov's file weighed heavily in Vasin's lap. The dead scientist's file picture was a professional studio portrait, the face cast in a dramatic half-shadow like that of a star from Mosfilm. Petrov wore his good looks lightly, a half smile on his lips. A face from a magazine: curly light hair, large blue eyes, a chiseled jawline. A face that nobody had smacked, certainly. The eyes ready to crinkle into an expression of earnest devotion. A lover's face.

Zaitsev and his men had been thorough. The file contained Petrov's complete personal records: forty pages of references and checks going back with clockwork regularity for each of the six years that he'd been in Arzamas. Party meetings attended and dues paid, formal reports from Party instructors. And before that his Young Communist League records and a pile of letters of recommendation from university supervisors. The letterheads bloomed with red stars and laurel wreaths.

Vasin's practiced eye caught what wasn't there. There were no denunciations from colleagues or snide notes from superiors in the file, no phone or mail intercepts. None of the usual fragments of office gossip or petty resentments that usually found their way into the *kontora*. The KGB, it seemed, had no eyes or ears inside the senior circles of the Citadel. As far as the *kontora* was concerned, the Institute was smoothly sealed behind a high, closed wall of silence.

It had been Major Efremov who had conducted most of the interviews with Petrov's colleagues in the days after his death. The language of the transcripts was a familiar high officialese, for the most part a dense wad of meaninglessness. But one witness stood out: Dr Vladimir Axelrod, Petrov's laboratory colleague and, by his own admission, personal friend.

EFREMOV O. P. (MAJOR, GUGB/AZ16): Comrade Doctor Axelrod, kindly present your estimation of the deceased's mental state in his final days.

AXELROD V. M.: My observation was that Dr PETROV exhibited no behavior that could be described as out of the ordinary.

Q: How frequently did circumstances afford you the opportunity to assess the mood and behavior of the deceased?

AXELROD V. M.: We saw each other on a daily basis when we were working on the same project. In the final days of his life this was the case. We also had frequent social intercourse with other comrades from the Institute.

Vasin rubbed his eyes and swore quietly. The formality of such records had always infuriated him, squeezing out the words' life and casting every subject into a predetermined role, the contrite criminal, the helpful citizen.

Q: Are you aware of any circumstances, professional or personal, that may have caused Comrade PETROV'S mind to be unusually stressed or disturbed?

AXELROD V. M.: We are all in a state of professional stress due to the urgency and importance of Project RDS-220.

On the page, fiercely typed out in triplicate, the investigator and his subject spoke like amateur actors declaiming lines from some archaic play. And yet, of all Petrov's colleagues, only Axelrod had been summoned for a second interrogation, this time with General Zaitsev personally. The tone of the next interview was more brutal. Zaitsev knew precisely what he needed from his subject.

> ZAITSEV O. V. (M-G GUGB/AZ16): Are you aware of any persons who may have introduced subversive influences into Dr PETROV'S life?
>
> AXELROD V. M.: I know of no such subversive influences.
>
> Q: He was known to read foreign literature of a nihilistic nature. Who pressed such psychologically unhealthy material on to PETROV?
>
> AXELROD V. M.: I know nothing of Dr PETROV'S literary influences. But he received his books in the same way as all of us. By special post from the Library of the Academy of Sciences, or from our institutes, or from our families. Most likely, PETROV'S father sent it to him. I suggest you ask the Comrade Academician. I understand that he is a man of wide interests.

A mistake on Axelrod's part, Vasin saw immediately, to try to crack a joke with a man like Zaitsev. He could imagine the General's meaty face flushing at the young man's insolence.

> Q: Answer the question. Did you supply him with any subversive or foreign literature?
>
> AXELROD V. M.: In July of this year I lent the deceased a copy of BEING AND NOTHINGNESS by JON POL SARTR (SP??), a French progressive.
>
> Q: And what is the nature of this book?
>
> AXELROD V. M.: It is a philosophical work of the existentialist school. The author asserts that an individual's existence is prior to his essence. He is concerned with proving that free will exists.

Q: Has this book been approved by competent authorities for the Soviet reader?

AXELROD V. M.: It is not a banned book, as far as I know.

Q: Answer the question. Has it been approved for general reading?

AXELROD V. M.: No.

Q: Because its content is subversive or anti-Soviet?

AXELROD V. M.: I cannot comment on what is and is not approved for the general reader or why. We are privileged here at Arzamas to have unrestricted access to foreign periodicals and literature because we need this material for our scientific work.

A spirited riposte. But between the lines Vasin's interrogator's eye could read, long before Axelrod saw it, the goal toward which Zaitsev's questioning was leading, steady as a tractor plowing a furrow.

Q: Did PETROV read much restricted foreign literature?

AXELROD V. M.: None of us have much time to read for leisure.

Q: Nonetheless, you would confirm that he was interested in such foreign philosophies? When he could, he read them?

AXELROD V. M.: He was interested.

Q: Therefore he was under the influence of the Frenchman SARTR (SP?)?

AXELROD V. M.: In a sense, yes.

Q: And you would confirm that in the weeks before his death PETROV'S social activities had reduced significantly?

AXELROD V. M.: All our activities have been significantly reduced by the RDS-220 program.

Q: But you confirm this to be PETROV'S case?

AXELROD V. M.: Yes.

Q: You can also confirm that he was showing signs of stress? Sleeping irregularly?

AXELROD V. M.: You could say the same for all of us at this time.

Vasin turned the page. Zaitsev had set up every part of his theory as carefully as a billiard trickster positioning his balls. Now he sank them, one by one.

Q: Comrade Doctor. You clearly failed to spot the signs of mental disintegration in your comrade in the days and weeks before his death. Do you feel remorse?

AXELROD V. M.: Naturally we all felt shock and remorse at Dr PETROV'S death.

Q: You personally felt remorse?

AXELROD V. M.: I felt remorse.

Q: You have confirmed that the deceased was under consider-ably increased work pressure. You have also said that his sleeping became erratic, his social life dwindled. Do you wish to deny or confirm these statements?

AXELROD V. M.: I confirm.

Q: Furthermore, you have said that PETROV was in the habit of studying foreign philosophical literature of a nihilistic nature that is not considered suitable reading for the gen-eral Soviet public. You have said that at least some of this material he received from you. You deny or confirm this?

AXELROD V. M.: I confirm.

Q: Would you accept the formulation that in your Commun-ist enthusiasm for your vitally important work for the Motherland you may have overlooked difficulties your Comrade PETROV was experiencing in his personal or interior life?

AXELROD V. M.: I accept it.

(Signed) AXELROD V. M.

(Signed/Interrogation conducted by) Major General ZAIT-SEV O. V.

(Signed/Interrogation witnessed by) Major EFREMOV O. P.

Vasin knew he wouldn't have to read Zaitsev's final report. The General's theory had been neatly laid out in the Axelrod interrogation. Petrov, in the official opinion of the KGB, had been driven to take eight thousand times the lethal dose of thallium by a deadly combination of overwork and French existentialism.

V

Vasin stretched wearily at his desk in the *kontora*'s registry. Outside the windows the light was draining from the gray sky. He closed the files and tapped his notes into a neat stack. He was hungry. But when he caught sight of Efremov bustling through the double doors with thunder in his face, he knew that the cafeteria would have to wait.

'Vasin?'

'That's me.'

'Listen, I don't know what you think you're playing at, or who it was that your people called, but . . .'

'The Professor is ready for me? That's what you came to say? You're kind.'

Vasin stood briskly, folded his papers into his tunic pocket, and smiled at Efremov's obvious discomfiture. A familiar enough scene: the local officer on the case realizing with ill grace that he's not master in his own house. He hadn't been expecting flowers. 'Never hurts to give the little jerks a little jerk,' Orlov had advised Vasin. 'They need to know who's in charge.' Orlov had himself in mind.

A *kontora* Volga sedan awaited them. The driver pulled a blithe U-turn across the central reservation and raced the car across the darkening town toward the Professor's home.

VI

The Adamovs lived in a handsome pre-Revolutionary building that overlooked the monastery on the far side of the Savva River. The facade was decorated with plaster caryatids and nymphs, their ancien régime voluptuousness rendered saggy by thick coats of Soviet paint. A single angry eye peeked from the concierge's booth in the hall, but was evidently satisfied by Vasin's and Efremov's uniforms. As they ascended the stone stairs, Vasin noticed that this place had none of an apartment building's usual cacophony: no noisy radios or raised voices, children's shrieks or slamming doors.

On the second floor, Efremov tugged a brass bellpull. After a long pause the heavy door swung open. At first Vasin thought that a boy had opened it. But it was a young woman with pale, close-cropped hair. She wore a pair of checked trousers, cut fashionably short, and a loose sweater. Her eyes were set wide apart, and her body was long-waisted and thin, like an elegant weasel's. She exuded an icy glamour.

The young woman leaned her head on the door and gripped the handle with both hands. She said nothing, but her eyes shone with an unnatural brightness.

'Excuse the disturbance. I believe I have an appointment with Professor Adamov?' Vasin stammered a little under the intensity of the young woman's stare. 'Major Alexander Vasin.'

'One second.' She spoke in a whisper. She crossed the wide hall-way, swaying unsteadily, leaving the two KGB men standing at the open door. He heard a murmur, and then Adamov's voice.

'Come.'

Adamov sat at the head of a dining table of dark wood surrounded by high-backed chairs with carved arms. The young woman took her place beside the Professor. An elegant, old-fashioned lamp that hung over the table illuminated empty plates and the two diners' hands, but left their faces in shadow. Adamov eyed his visitors with unconcealed distaste.

40

'Comrade Majors. Sit.'

Vasin took the chair to Adamov's left, leaving Efremov to settle at the far end of the table. Vasin felt that he had wandered into some kind of interrogation scene from a historical film. In the half-light, Adamov's face looked cadaverous. Next to him the girl sat poised and motionless, as if posing for a portrait.

'Professor Adamov, thank you for seeing me. It has also been explained to me very clearly that your work is of the utmost national importance.'

'I protest this waste of my time, especially at this critical juncture in the fate of our nation. But when I receive a call from a member of the Politburo, I have no choice but to obey. So. Quickly. Fyodor Petrov.'

'Exactly, sir. Was there anything about his behavior in his last days that seemed strange to you? Did you notice any signs of distress?'

'I noticed nothing amiss. I have already spoken of this to your colleague. Him.'

Adamov gestured down the table at Efremov as though indicating an inanimate object.

'I have read your statement. Perhaps you would describe your relationship with Petrov to me?'

'Petrov was one of my most promising assistants. He had a good mind. Our relationship was perfectly correct and professional. He will be missed.'

'So Petrov was generally well liked?'

A pause.

'Everybody loved Fedya.' The young woman uncurled herself and leaned forward into the light. Her voice was low and slurred. 'Everyone. Just. Loved. Fedya Petrov. Especially my *husband*.'

Tension snapped like an electric spark down the table. The woman was young enough to be the Professor's daughter. Vasin watched her face tighten into a little smile. Mascara was starting to run from one eye. The whine of a boiling kettle rose from the kitchen.

41

'Maria.' Adamov spoke firmly, as if to a child. 'Would you bring us some tea, please?'

She stood, abruptly, and stalked out of the room.

Adamov turned back slowly to Vasin.

'Continue.'

'How long did you know Petrov?'

'Have you spoken to the boy's father? Our esteemed Comrade Academician Arkady Vasilyevich Petrov?'

'I have, sir. I spoke to the Academician two days ago in Moscow.'

Vasin thought of Petrov senior, slumped and weeping under his dripping dacha eaves a few days before. A plump man, punctured by grief.

'So doubtless Arkady Vasilyevich told you that we have known each other a long time. Since Fyodor was a boy, in fact.' A tremor entered Adamov's voice. 'You see, his father and I were colleagues, once. Back in the thirties. The heroic days.'

His voice was bone-dry, like papers being taken down from dusty shelves.

'Are you still on good terms with Academician Petrov?'

'Very good.'

'And his son Fyodor Arkadiyevich came here to work for you . . .'

'In 1955. You will surely see that in the files. Please, Major, let us spare each other the pro forma questions.'

Maria returned with a tray laden with clinking china. It seemed to take all her concentration to pour out three cups of weak tea. She passed them to Vasin, Efremov and Adamov with great formality, then resumed her place in silence.

'Tell me about the safety procedures in your laboratories.'

'Speak to Dr Vladimir Axelrod, if you must. He is aware of the technical aspects of the work he did with Petrov. He will be at his post tomorrow.'

'The pathologist expressed doubt that Dr Petrov could have received such a large dose by accident. Not in the laboratory.'

'I defer to his opinion.'

'But if the doctor's estimation is correct and it was not an acci-
dental poisoning . . .'

'Then the poor soul knew what he was doing.'

'Or someone gave it to him,' said Vasin.

Adamov's face did not flicker.

'You are saying someone in this city could be a murderer?'

'I am saying that someone in your laboratory could be a
murderer.'

'The stoker sees other stokers everywhere.' Adamov pronounced
the old Russian proverb in an indifferent voice. 'An investigator, I
imagine, sees murderers everywhere. I pray you are wrong. But the
fact is, Comrade Major, that with the work we are undertaking at the
laboratory, nobody has time to pursue your theory. Project RDS-220
is too important to be interfered with. Eight days from now the most
powerful bomb the world has ever known will be detonated. That is
all you need to know. So let me put it more plainly. Objectively, I
cannot afford to give a flying motherfuck how Petrov died.'

Adamov pronounced the words precisely. In the Professor's
clipped voice the profanity was as shocking as a bucket of turds
tipped onto the white tablecloth. Vasin was stunned into silence.

'I mean you no disrespect, Major,' Adamov continued smoothly.
'Indeed it is quite possible that you are an intelligent young man.
You show some signs. You are courteous, certainly, which is not the
case with many of your colleagues. But please. File your report.
Allow us to work in peace.'

'Professor, I care what happened to Petrov.'

Adamov sipped his tea. The four of them sat in silence for a long
moment. Vasin could sense Efremov's pent-up anger radiating down
the table like heat from a stove, but refused to catch his colleague's eye.

'Why?' Maria's voice cut abruptly across the room, slightly
slurred. '*Why* do you care, Comrade Major?'

Vasin could barely see her shadowed face. Could this girl be the
professor's wife?

'Comrade . . . Adamova? Because we cannot live by lies.'

'I see,' said Adamov. 'You are a believer in General Secretary

43

Khrushchev's brave new world. The end of Comrade Stalin's personality cult. A new heaven and a new earth. Very laudable.'

Maria gave a soft snort.

'Can it be that we have lived to see the day?' she drawled. 'An officer of State Security who tells us about truth.'

'Masha. Enough.'

She leaned forward once more into the light. Vasin had seen drunken bravado before. But Maria Adamova was different. Her green eyes shone with an almost supernatural intensity. She took a breath, as though to continue, but her husband interrupted.

'We are all tired, I think.'

'A final question, Professor. When was the last time you saw the deceased?'

Adamov drained his teacup before answering.

'You are playing games now, Major. You know the answer already. We saw Petrov the evening before he was taken ill. He came here for dinner with Colonel Korin. We discussed the project, as usual. There was nothing out of the ordinary. Korin left at ten to catch a flight north, to Olenya. Petrov sat with us a little longer. We debated some technical issues. He appeared tired, but resolute. As we all are. Now, you must excuse us. Maria Vladimirovna will see you out.'

Vasin stood and shook Adamov's papery hand. Masha walked them to the door, moving slowly as though with infinite weariness.

'My thanks for your time, Comrade Adamova.'

Maria's eyes wandered slowly around Vasin's face.

'How fortunate to be guarded by honest men.'

The door shut behind him with a heavy thud.

VII

Vasin and Efremov stepped out into the empty street. A shower had passed, and the wet tarmac shone in the streetlights like a policeman's plastic cape. A wintry smell rose off the bare earth of the municipal flower beds and freshly planted linden trees on the avenues. In the waiting car the driver's face was illuminated by the pale yellow light of the dashboard.

'Satisfied, Vasin?'

'Satisfied. Thanks for the lift, Efremov.'

The adjutant opened the car door and motioned Vasin inside.

'Think I'll go home on foot. Clear my head.'

Vasin turned so that the wind would be at his back and began walking before Efremov could stop him. The domestic hour, after dinner and *Good Night, Children* on the television. The hour of tea and vodka, arguments and lovemaking. The windows of most of the apartments were illuminated with warm yellow light.

In Moscow, Vera would be on the phone to her girlfriends. Vasin could imagine his wife's gossipy voice, see the thin stream of cigarette smoke curling out of the kitchen window. Nikita would be sleeping, calm, in his narrow bed under the bookshelf. By day the boy's face was usually anxious. Admonitions, scolding, advice, the poor kid lived his life advancing doggedly into a daily blizzard of instruction. Only when he slept did his features relax. Nikita was an obliging child, eager to please. But there was no pleasing Vera. 'Next time you will do even better,' she would tell the boy. 'Higher, higher, and ever higher!' Somewhere else in the world, she believed, there was always a child who was better than her own.

Vasin's thoughts paced in cramped circles. Arguing voices, call and answer, like a ritual song. Guilty, Vasin stood in the center of a wheeling parade of sins summoned by his wife Vera's rage. Here, the vodka. There, his mistress.

'How could you?' Vera had screamed. 'With *her*?'

From the next room, the thump of piano scales as Nikita desperately hammered arpeggios to drown his parents' arguing voices.

'You *bastard*!' Vera had added, perhaps for the neighbors' benefit.

Out in the world, criminals, hard men, begged for Vasin's mercy. In his own home, he hung his head like a defendant before a People's Court, searching the parquet in vain for crumbs of forgiveness.

Vasin, he thought, your life is ridiculous.

He reached the end of a long boulevard. Like all the streets of Arzamas, it seemed to terminate in a park. Beyond, he guessed, was the forbidden perimeter with its guard towers, dogs, and barbed wire that he had crossed by train the previous day.

Vasin was ravenous. At a tiny cafeteria by the train station, he bought himself a plate of sausages and a mug of watery beer. The other customers were workingmen in overalls and greasy caps, but the place exuded none of the underworld squalor of similar establishments in Moscow. Vasin leafed through General Zaitsev's copy of *Krokodil* as he stood by his tall, rickety table. A witty dispatch from the Kharkov Tractor Works, whose male-voice choir had just won an all-Union singing competition. By half past ten he was the last customer. The burly waitress began scooping the remaining sausages out of their steaming water and tipping them, wriggling like live things, into a jar. Her own family's dinner, doubtless.

'We're closing, Comrades,' she chirped to the empty room.

'Tell me, pretty one,' Vasin called, raising his voice against the clatter of pans and swish of water. 'Where does a man go to get a drink at this time of night?'

'The Café Kino, of course.' The woman looked him up and down in frank appraisal and cocked a flirtatious eyebrow. 'But you should watch out. There might be pretty girls there.'

Vasin badly wanted to speak to somebody. A stranger would do. A stranger would be better, in fact. And he needed a drink.

'You've earned it, you brave boy.' Katya Orlova's words, as she sloshed out brandy from her husband's crystal decanter. He thought of her pendulous breasts, her rouged mouth, her desperate sexual hunger.

What the hell were you thinking? The boss's *wife*?

Vasin faced the spitting wind and retraced his steps toward Lenin Square. In front of the glass facade of the Kino Moskva the tramlines gleamed. A light burned in the vestibule, and from a basement came the faint sound of chatter and music. There would be cognac. And girls. This, thought Vasin, could end badly. He went in.

The Café Kino occupied a large, dimly lit basement. In one corner a dozen young people had pulled chairs into a circle and were talking loudly over the din of swooping, rhythmic music. It sounded like – could it be? – American rock-and-roll. Thrilling. Semilegal. A couple of the customers glanced at Vasin's uniform as he hung his mackintosh and cap. Vasin saw no fear in their eyes, only distaste. He settled onto a stool at the long bar and ordered Armenian cognac.

'Unusual music.'

The barman was an indigenous Siberian. His flat, Oriental face was expressionless.

'*Rei Charlz*,' he said. '"*Hit road, Dzak*". The kids bring their records in. *Mo-town*.'

In Moscow, Vasin had seen bootleg copies of foreign discs cut onto the celluloid of old X-ray sheets. 'Bone discs,' the young people called them. Each cost a month's student stipend. But the records scattered over the far end of the Kino's bar were originals, in brightly colored sleeves that spoke of America and unattainable luxury.

A couple stood to dance in front of a small, empty stage. The girl's hair was done up in a beehive and the young man's was glossy with cream. They performed a kind of half-squatting dance.

'"*Da Tvist*",' explained the barman, unprompted, as a new song came on. '*Chabi Cheka*.'

He pronounced the last word like 'che-ka,' the first Bolshevik secret police, now slang for the KGB. Was the man being sarcastic? But the barman's face was blank. Vasin took his cognac to a table in an empty corner.

Pizhony, they called these kids in Moscow. The stylish ones. The term mixed contempt and envy. Plain working people would travel specially to gawp at the *pizhony* preening up and down Gorky Street in their polka-dot dresses and sharp suits on a Saturday night. Vera hated them. 'Today he dances jazz,' she quoted from a *Pravda* editorial. 'But tomorrow he will sell his homeland.'

So this was Petrov's world. French books, American Motown music, *pizhony* for friends. The handsome only son of Academician

Petrov had moved from the bubble world of Politburo compounds around the dacha village of Zhukovka to the still more isolated cloud dwelling of Arzamas. When he died, the young man had left a great future behind him.

Very senior people were taking a close interest in the circumstances of Fyodor's untimely death. General Orlov had spelled that much out in his cluttered Moscow office.

'We have made a promise to the Comrade Academician to get to the truth of the matter,' Orlov had said, his face cracking into a toadlike grin. 'I told him we would put one of our best men on the case. Our very best man.'

Orlov was stating a fact. Vasin was one of the best investigators in the *kontora*. He knew this because he had been so hated for so long. In his ten years at police headquarters, Vasin had acquired a reputation for maddening tenacity. A towering sense of righteousness inherited from his mother. A faith in science from his father. Put together they had made Vasin a brilliant detective – as well as a giant pain in his colleagues' backsides. Honesty was an unusual quality in a police officer, and certainly not a career-advancing one. At least until Vasin's work had caught the watchful eye of the *kontora*.

General Orlov, in Vasin's first ever interview at the Lubyanka, had put it bluntly.

'Too many people in this building are skilled in covering their asses, Vasin. Spinning fairy tales they think the bosses want to hear.' Orlov's jowls bulged over his uniform collar; his small black eyes skewered Vasin like pins in a butterfly. 'You're a man who can actually investigate a crime. And sometimes we need the whole truth.'

Vasin had noted the *sometimes*.

Special Cases was Orlov's term for his tiny, secretive department, which occupied a suite of shabby offices on the Lubyanka's ninth floor.

'Special cases concern people who must be treated with special sensitivity,' Orlov had explained the first time Vasin brought him a completed case file, ready for the Prosecutor's Office, and locked all three copies away in his personal safe. The General had turned full face to the indignation kindling in Vasin's eyes.

'You understand, Comrade Major?'

'Of course, Comrade General.'

Orlov settled back into his chair.

'Agreement that comes too easily is usually not agreement, Comrade.'

Orlov sized Vasin up like a newly caught specimen.

'Ah! You say nothing!' The General pointed his index finger right down his line of sight into Vasin's face. 'Good! You do not rush to assure me that yes, Comrade General, you do most certainly agree with my every word. Because you do not agree. So, good. You learn fast, Vasin.'

'Sir, I . . .'

'You believe that a criminal's place is in jail. Am I right? Of course I am right. For every crime committed, a criminal must be punished. That is the simple arithmetic of our police comrades. And you will agree that they are simple men. "Vasin?" your bosses said. "Vasin is the cleverest one." They were not paying you a compliment. But you are clever. So I invite you to consider a different logic. Here, we have crime.'

Orlov, sliding forward on the black leather of his chair, picked up a glass paperweight that contained an outsize dragonfly and placed it to one side of his blotter. 'And here, punishment.' A heavy-lidded malachite inkwell stood in for the Gulag.

'Your old job was simple. Connect our guilty four-winged comrade here' – Orlov's fingers mimicked a little man wandering in zigzags across the spotted paper – 'and place him in here.' He snapped the top of the inkwell shut.

'A crime must be punished. But what do we say about actions that are no longer crimes? Or not yet crimes? What do we say if the criminal is doing important work for our Motherland? Do we follow justice and shut him up – even if it damages the cause of international Socialism? Or let's say, we leave a criminal at liberty in order to catch a bigger criminal. Or, perhaps there is a higher motive behind the crime? As you surely know, Comrade Stalin was a bank robber once, in the service of the Party. Would you wish him to have been sent to the tsarist gallows? Which part of you wins the debate, the good Communist or the good policeman? I see you catch my drift. Special cases. They need a special kind of investigator.'

Vasin had made his face stone. Orlov's scrutiny was like a beam of cold light.

'You know, young Sasha, my father was a priest.' Vasin shuddered involuntarily from the double shock of Orlov's abrupt familiarity and his admission of a background most men would go to great lengths to hide. 'His father was before him. Generations of Orlovs. Humble parish priests, dispensing opium to the masses, soothing them with lies and scented smoke. And I, too, was educated at a seminary. Just like Comrade Stalin. Did you know that, about either of us? In my early days in the Cheka my comrades would call me *pop*, the priest. There were actually quite a few of us in the *kontora*, back in those days. Priests, I mean. Sons of priests. Also lots of bitter Jews, revenging themselves on the world. Georgians, too, of course. How we underestimated them! Anyway. Here is what every priest learns as he hears men's confessions. In order to do evil, a man must tell himself he is doing good. When they kneel and whisper and cross themselves they say, "I ask forgiveness." But in truth they are asking for understanding. *Then the Lord saw that the wickedness of man was great on the earth, and that every intent of the thoughts of his heart was only evil continually.* Ever heard that? Of course you haven't. The Bible. Book of Genesis. Now our respected Comrade Dzerzhinsky also thought all men were evil. Perhaps he was right. But the important thing is that every man I have ever met secretly thinks that he is better than all others. More deserving. He wants the world to recognize his worth. There is no righteousness and no wickedness. Only men, imposing themselves on the world, for reasons they consider good. And justice exists no more than God does. What we humans – even enlightened Soviet men – call justice is just another name for expediency. The moment you understand that is the moment that you become an investigator worthy of Special Cases.'

There was a pause as Orlov moved the dragonfly and the inkwell carefully back to their places.

'You may rely on me entirely, Comrade General,' Vasin said to Orlov's face, saluting.

Special fucking *criminals*, he'd said to himself on the way out.

Vera loved the perks. The new apartment. A private phone line. A place at the best Young Pioneer camp for Nikita. A Moskvich car of their own, one day. Jolly drinks parties at the officers' clubs that now occupied the palaces of Moscow's old merchant princes. Crystal glasses, white-coated waiters, generals and colonels in boots that other men had shined for them. Glittering women. General Orlov's wife like a galleon in full sail, bursting from her evening dress. Her confiding hand on the jacket of Vasin's new uniform, her eyes wet with lust.

The cognac was sweet. Vasin ordered another. The barman brought over the bottle and left it on the table. He was in a city that existed on no maps and had no address other than a post-box number. A city where Armageddon's engineers danced to Negro music. Once again trying to discover the misadventures that had led a Party princeling to an early grave.

'Because I care,' Vasin heard himself telling Adamov. 'Live not by lies.' You pompous idiot. Vasin could feel the brandy thudding in his temples. You're just a janitor. Bring me your embarrassments, your peccadilloes, your jealousies, and your addictions. And Vasin will mop them all up into a slim file and take it to Orlov for discreet burial in his steel safe. And in the afternoons he will screw the tits off the General's needy wife.

Live not by lies.

He ran a hand through his hair and swung a heavy gaze over the bodies of the girls in the corner. Perhaps he could join them? Ask if they knew Petrov? Suggest a dance.

Oh for God's sake.

From behind Vasin came the scrape of a chair. Kuznetsov, his oiled hair falling over his brow in an untidy lick, settled his sturdy bulk at Vasin's table.

'Enjoying our Arzamas nightlife? Hear you've been busy.'

Vasin, slowed by the drink, focused blearily on his roommate. To his own surprise, Vasin found himself pleased to see him.

'Help me finish the bottle?'

'It's in my job description, old man. Duty to my Motherland.'

Vasin was back under the *kontora*'s watchful eye.

CHAPTER THREE

MONDAY, 23 OCTOBER 1961

SEVEN DAYS BEFORE THE TEST

I

Vasin awoke late. He found Kuznetsov in the kitchen, immersed in a copy of *Science and Life* magazine.

'Morning. Coffee's cold, I'm afraid.'

'It's nearly nine. Why didn't you wake me?'

'Thought you needed your beauty sleep, old man. Seemed tired last night.' Kuznetsov looked up from his magazine with an affectionate smile. 'We want you to have a restful time in Arzamas.'

'Fuck off, Kuznetsov.'

'You're welcome.'

Vasin dressed quickly in his civilian clothes, swigged the remains of Kuznetsov's cold coffee straight from the pan. There was something about last night's conversation with Adamov that had sent Vasin's mind running. That catch in the Professor's voice as he spoke of Petrov's father. A curious tension as he spoke of his old colleague from the heroic days.

'Drive me to *kontora* in that crate of yours?'

'Your wish is my command, O master.'

'And while I'm there, I need to speak to Axelrod.'

'Axelwho?'

'Come on. Vladimir Axelrod. Petrov's colleague from the lab. Adamov said I should speak to him about laboratory security.'

'*Adamov* said that?' The warmth had vanished from Kuznetsov's voice.

Vasin paused as he pulled on his mackintosh and shot his room-mate a raised-eyebrow glance.

'Want me to send some more telegrams? I thought we were all in a hurry to get this done.'

'Okay, okay. Calm down. I get it, old man. You can summon lightning from the heavens. No need to repeat the trick. I'll find you your Axelrod. Will have him washed and brought to your tent.'

II

The registry of the KGB headquarters was almost deserted. The duty archivist, a plump brunette with a lopsided hairdo, looked up resentfully from her novel as Vasin leaned into the doorway.

'Registry's shut till lunch. Sanitary day.'

'Library?'

'Library's open.'

'Library's what I need. *Great Soviet Encyclopedia*?'

Vasin tried a smile, which the girl extinguished with an irritated sneer.

'Over there. General Reference.'

He recognized the familiar dull red volumes massed officiously in the corner.

'Thanks. Here.' Vasin pulled Zaitsev's copy of *Krokodil* out of his mackintosh pocket and tossed it onto the girl's desk. 'It's a good one. Hilarious piece on collecting the sperm of champion bulls.'

Vasin quickly found the encyclopedia entry on Fyodor Petrov's father.

'Petrov, Arkady Vasilyevich, born St Petersburg, Russian Empire, 10 July 1901. Nuclear physicist. Member of the All-Union Academy of Sciences. Hero of Socialist Labor. Laureate of the Stalin Prize, 1951.'

Arkady Petrov's brilliant career occupied half a page. A doctorate at age twenty-one from the Physics Department of Leningrad State University. Work at Niels Bohr's Institute for Theoretical Physics in Copenhagen, 1930. Then Cambridge in 1934, working with Pyotr Kapitsa. Zurich, 1935. Returned to the USSR 1936, appointed deputy head of the Department of Theoretical Physics at the Institute of Physics and Technology in Kharkov, Ukraine. Became the department's director the following year. Then a place at the All-Union Academy of Sciences; prizes; State laurels.

Vasin puzzled over the list of Petrov senior's discoveries and scientific articles: 'The Density Matrix Method in Quantum Mechanics.' 'The Quantum Mechanical Theory of Diamagnetism.' 'The Theory of Superfluidity.' He understood not a word. Only at the very end of the entry did Petrov's official life revert to simple human terms: 'Spouse: Nina Petrovna Scherbakova, b. Kursk 1913, d. Moscow 1959. Issue: Fyodor Arkadyevich Petrov, b. Moscow 1929.' In the encyclopedia's next edition, doubtless, the painstaking editors would add the date of Fyodor's death.

Vasin opened another volume.

'Adamov, Yury Vladimirovich. Born Baku, Russian Empire, 9 December 1900.' Adamov's path through the Soviet physics world had been no less stellar than Petrov's. A post at the People's Commissariat for Education. Work at Göttingen and Leipzig in the twenties. He had, according to the encyclopedia, put his name to the 'Adamov pole in quantum electrodynamics', 'Adamov's equations for S matrix singularities', 'Adamov's theory of second-order phase transitions'. Whatever those might be.

Then Vasin spotted it. '1932–37: Head of the Department of Theoretical Physics, Kharkov Institute of Physics and Technology.' So Adamov had been Petrov senior's boss in Kharkov, before Petrov took over his post.

But what followed immediately after was stranger. '1944: Senior researcher, All-Union Scientific Research Institute of Experimental Physics.' Nothing between 1937 and 1944. Petrov's entry for the same period had been a steady stream of publications, academic posts, prizes. But for seven years Adamov's biography was a perfect blank. Which could mean only one thing.

The Professor had spent those years as a prisoner in the Gulag.

III

By daylight the colonnaded entrance of the Citadel, with its crowds of hurrying people, reminded Vasin more than ever of a grand transport terminus. And like a railway station, the facade on Kurchatov Square concealed a deep hinterland of offices, glass-roofed laboratories, and bunkers that extended back into the far distance, shut off from the surrounding boulevards by high brick walls. Kuznetsov effortfully parked his car almost straight in an official spot under the critical gaze of a traffic policeman.

Vasin recognized Axelrod from the crowd of assistants around Adamov the night of the lecture. He was angular and pale, with the kind of face that was sharp in the flesh but fudged in memory. He stood alone in the echoing lobby, his clothes hanging off his skinny frame.

'Axelrod. Vladimir Moiseyevich.'

The scientist pronounced his name in a murmur, standing almost to attention.

'Vasin. Alexander Ilyich.'

They shook hands. Vasin removed his trilby hat and tried a smile that was not returned. Axelrod peered through rimless spectacles, waiting for his visitor to speak.

'Thank you for seeing me at such short notice. I know how valuable your time is.'

Axelrod's faced twitched at the irony. Did he have a choice?

Axelrod motioned the KGB men toward the turnstiles. Kuznetsov made to go through, but Vasin stopped him with a hand on the shoulder, leaning close to whisper into his ear.

'Can I handle this one alone? He looks like a nervous customer to me.'

Kuznetsov began to answer, but Vasin cut him off.

'Please. Trust an old detective. It will save time, I promise.'

Kuznetsov frowned, but nodded.

After their passes had been scrutinized, Axelrod moved quickly, disappearing at speed down a corridor that led into the heart of the building.

They reached a stairwell fragrant with tobacco smoke.

'Wait. Vladimir Moiseyevich,' called Vasin. 'Shall we?'

Axelrod resignedly retraced his steps. Vasin offered him an Orbita, but the scientist declined in favor of his own, an unfamiliar blue packet with a swirl of stylized blue smoke on the cover. Foreign.

'You know why I am here?'

Axelrod nodded.

'You were friends with the deceased?'

'Yes. Dr Petrov and I were friendly.'

The man was clearly nervous.

'I saw your late friend yesterday morning.'

Axelrod went pale.

'What do you mean?'

'In the morgue.' Vasin paused a moment to let the thought sink in. 'What killed him was terrible.'

'I . . . I can't imagine.'

'I think you can imagine it. And I believe it keeps you awake at night.'

Axelrod said nothing, inhaling deeply through his cigarette.

'I've read your interview. General Zaitsev is a brute.'

A spark kindled in Axelrod's eye. A sideways glance that said: Come now, Major, you're going to have to try harder than that.

Vasin plowed on.

'Understand Zaitsev. To an axe, every problem looks like a log to be split. Zaitsev's job is to provide solutions that are acceptable to all. So, he looks for a solution to the Petrov problem. Negligence? The Institute will be upset. Madness? Petrov's father won't believe it. So Petrov himself cannot be the culprit. So Zaitsev finds other factors that can be blamed. Subversive foreign literature. Pressure of work. Comrades too distracted by their heroic labors to notice a mind under strain. In Zaitsev's world, he begins with the solution and works back to his suspect.'

The scientist looked Vasin over with a new curiosity.

'Are you always so frank with your interviewees?'

'No. But they're usually not cultured men such as yourself. I thought we would save time by dispensing with the formalities.'

Axelrod brushed the compliment aside.

'By saying this you wish to tell me that Zaitsev's world is not your world?'

'I am saying that I wish to get to the truth.'

'Of course you do.'

Axelrod sounded unconvinced. He volunteered nothing else, and they crushed out their cigarettes in unison. The urn had been freshly emptied, Vasin noticed, not the usual filthy mess of week-old butts he found in most offices.

'You wanted to inspect Dr Petrov's last workstation, Comrade Major?'

Vasin followed Axelrod's retreating figure as he clattered down stairs, deep into the basement. The levels were marked with codes. The young scientist finally stopped before a set of heavy double doors marked LABORATORY ZH-4.

They stepped inside a vast concrete hangar lit with rows of industrial lamps. In the center was a machine at least as long as a locomotive, a massive tubular structure sprouting cables. On either side was a row of steel-cased instruments with buttons and dials that reminded Vasin of soda-vending machines. The place had the sacramental atmosphere of the Kremlin cathedrals he had visited with Nikita, cavernous and full of unfathomable mysteries. He sniffed the air. The place had a familiar odor that Vasin couldn't place.

'The smell? Everyone notices it the first time they come in here. It's the same as the metro. High-voltage electricity creates charged plasma when it passes through air. Ionizes it, creates ozone. That's what you're smelling.'

Axelrod led the way past banks of instruments, some attended by lab-coated young men who cast quick, unfriendly glances at Vasin as he passed. Raised voices came echoing down the hall in a high-pitched argument. Axelrod paused for a moment to tune in to the quarrel, got the gist, then dismissed it with a shake of the head.

'Sorry. Everyone's on edge around here. Not much sleep. The test, you know. Working the machine day and night.'

They had paused at the foot of the giant cylinder with a clear view along its whole mighty bulk, twice the length of a metro carriage.

'What does it . . . do?'

Axelrod grinned for the first time. Here he was on home territory.

'It's a mass spectrometer.' Axelrod registered the investigator's blank look. 'We bombard samples with ions and pass them through a magnetic field. It separates the stream into their different atoms, by weight. The heaviest atoms bend more. This one is based on an American design they called a calutron. It's mostly used for separating uranium into its different isotopes. On an industrial scale, it's one of the ways of refining uranium to weapons grade. But this is a baby one. For experiments.'

As a young detective Vasin had wasted hours following up citizens' paranoid denunciations of neighbors who were taking too close an interest in radio parts. But here Axelrod scattered military secrets like the husks of sunflower seeds.

'Don't look so alarmed, Major. There's nothing secret about the principle. While we were having our Great October Revolution, the English were building the first of these machines. In forty years' time, I promise you, the capitalists will be trying to copy ours.'

A row of large glass-walled rooms lined one wall. In each one stood a laboratory bench topped with powerful ventilator fans. Axelrod led Vasin up to one of the plate-glass windows.

'This is where the samples are prepared for spectrum analysis. They must be turned into gases for the process to work. And – there – you see where the samples are kept.'

59

Vasin followed his guide's pointing finger and saw a glass cabinet in the corner of the chamber. It had its own ventilator and contained stacks of dull metal cylinders the size of large coffee mugs. In the corner stood an instrument that Vasin was coming to recognize, a Geiger counter.

'The radioactive samples are kept in those lead cylinders. One or two grams in each.'

They were interrupted by a momentary dimming of the lights in the hall, like the lamps on the metro signaling the arrival of the last train. A bass thrum began to vibrate through the concrete floor.

'Bother,' said Axelrod.

Bother? Vasin looked askance. That's how they swear round here?

'It's the next-door laboratory. We share a generator. They're starting up their pneumatic rams. They're meant to bloody warn us.'

'Pneumatic rams?'

'For a barometric chamber. Measures sudden changes in pressure. We built it for particle and gas research. But for the last year they've been letting Dr Mueller use it. One of our German guests. Been with us since the war. He uses it for his experiments on shock-wave effects.'

The lights flickered slightly as the thrum increased. Vasin turned back to the workstations.

'Petrov worked here?'

'Right here.'

'And he used thallium?'

'Certainly. We use thallium as a control. It's one of our most predictable alpha emitters.'

'Alpha?'

'There are three general types of radiation. Gamma is the most penetrating. Cosmic rays are gamma radiation. They cross the universe at the speed of light and travel right through the earth. Thousands of them are passing through your body every second. The atmosphere screens us from their most harmful effects. Beta radiation is less penetrating, but more dangerous because it has more energy. Alpha radiation is the opposite of gamma. It has far more energy than beta or gamma. But luckily for us it cannot penetrate human skin. Thallium emits mostly alpha radiation.'

'So why . . . ?'

'Why is it deadly? Because once it gets inside the body it is absorbed by the digestive system. Your bloodstream takes it to every part of the body. The alpha particles destroy everything around them, especially human cells. So every organ disintegrates from inside. The effect is . . . as you saw.'

Axelrod's face flushed, like a child's, and tears abruptly spilled from his eyes.

The scientist turned and walked quickly away, a handkerchief to his face. Vasin tactfully turned his back. The cellar looked like a cave from a futuristic fairy tale. A captive monster squatted obscenely in its center sprouting wires from its head and guts. To dominate and conquer one's fellow man, that was easy. The stupidest and most brutal of Vasin's colleagues could do that in minutes. But to enslave nature itself to your purposes, like a genie inside a lamp? That was something akin to black magic.

'Forgive me.' Axelrod returned, his freckled face blotched red.

Vasin nodded in sympathy. The shock wave of bereavement could sometimes be an ally, he knew. He'd seen it burst men's composure apart at the seams. Made them want to confide in strangers. He was glad of Kuznetsov's absence.

'Do you have any idea how a man in this laboratory could ingest thallium by accident?'

'No.' Axelrod's voice had become a bleak whisper. 'In theory it might be possible that a few milligrams might escape from the fume chamber. We wear gas masks in there, of course. But we swabbed the one that Fyodor was using the day before he was taken ill and found nothing. He was a first-rate scientist, Major. And the technicians who work here designed the fume chamber. They would know if someone had made a foolish mistake. Not to mention the radiation alarms, those boxes that look like gramophone speakers along the wall. So, no. To answer your question, I do not see how a man could ingest a fatal dose of reagent by accident and not know it. But it's hard to say without knowing how much was inside him. No one told us. Or me, anyway.'

'The pathologist believes Dr Petrov ingested a lot.'

'How much?'

'Two thousand milligrams.'

Axelrod blanched.

'God almighty. He said milligrams? Are you sure? He must have said micrograms.'

'No. I know what he said, Doctor.'

'Two grams . . .' Axelrod barely registered Vasin as his eyes darted to the fume chamber, to his white-coated colleagues huddling over a disemboweled meter. He lowered his voice still further, making Vasin lean in to hear. 'Two grams is the weight of an entire sample. We use maybe a thousandth of that as a control in the spectrometer. That is not an amount that could in any way be taken accidentally.'

Axelrod looked down.

'The *kontora* says that Petrov signed for and removed two thousand milligrams from the lab himself,' continued Vasin. 'He stockpiled it bit by bit, for over a month. Signed for it, but didn't use it. Then he ingested it.'

'Two thousand milligrams unaccounted for? That's what they're saying?'

'It's in the experimental logs. You seem surprised.'

'That's not possible.'

'What part? Technically or personally impossible?'

'Both.'

'How do you keep track of the thallium you use?'

'Petrov had been running his own experiments in the calutron for weeks. Using thallium, naturally. They're all in the results log. He was testing different alloys of unenriched uranium metal. Different casings for RDS-220.'

Axelrod gestured down the row of fume chambers, each with its stack of lead pots.

'We know what comes into the laboratory. The senior technician takes delivery of dozens of capsules of material every shift. Then each laboratory chief records what is used in every experiment. You could compare the two to check if there was a discrepancy. But it would take days.'

'That's exactly what the *kontora* must have done. They said it was a huge job.'

'But *thallium* is very unstable. Half-life of just seventy-three hours. That means that half of it decays into mercury every three days. In six days, only a quarter is left. And so on. Makes it hard to keep track of.'

'You're saying this stuff just vanishes over time?'

'Something like that.'

'So the lab records . . .'

'Must be falsified. Those two thousand milligrams you say he supposedly signed for? In three days they'd have decayed to a thousand. In twelve, there'd be one hundred and twenty-five milligrams left. In thirty days, less than two. This story of him signing out thallium over a *month*? A piece of nonsense concocted for scientific illiterates. Show me that log, and I'll prove it's been faked.'

Vasin paused as the implications of what Axelrod had told him sank in.

'And you suggested a moment ago that it was *personally* impossible for him to have taken it himself? You knew him well. Did Dr Petrov show any signs of being suicidal? Had anything happened to upset him?'

Axelrod shook his head.

'No.'

'No, he didn't show any signs? Or no, you don't want to tell me?'

'He was not suicidal. Tired, perhaps. The project was everything to him. Like all of us, he was nervous about the test. He was looking forward so much to seeing it. But who knows what is really happening in a man's mind?'

Vasin nodded in understanding.

'And his relationships? Was he involved with anyone?'

'No.'

'Was he close to Maria Adamova?'

Other investigators swore that they could tell in an instant when a witness was lying. Vasin had seen too many honest men scared and flailing for salvation to make such a claim. But he could at least tell when a man was collecting himself. A contraction of the mouth. Shoulders tightening. The mouth drying into a tight smile.

'I don't know.'

Abruptly the ambient vibration stopped and was replaced by the distant sound of an electric klaxon, hooting.

'What the hell is that?' asked Vasin in alarm.

'Here we go.' Axelrod winced. 'Wait for it.'

A powerful thud echoed through the floor, like a train colliding with a concrete wall.

'Dr Mueller's shock waves.'

Axelrod led Vasin out of the spectrometer hall. In the corridor they stood aside to allow a pair of technicians to pass, wheeling a large trolley from the neighboring laboratory. Vasin glanced inside and saw a pile of shit-stained fur, goats' horns, some hooves sticking up at odd angles. The animals had been crushed, as though stamped on by a giant boot. A farmyard reek, utterly incongruous in this sterile place, followed them.

'Ah. And here is our German comrade himself. Good day, Dr Mueller.'

Mueller, a small man in round glasses, blinked nervously as though expecting to be slapped. He nodded briefly to acknowledge Axelrod's greeting and scuttled on. With a bang of doors, the German and his trolley were gone.

Axelrod waited before answering Vasin's unasked question.

'They found him in one of the German camps. He'd set up a whole laboratory for medical experiments. Used the prisoners as his guinea pigs. The Americans wanted him hanged, but we decided to use his bright mind. So here we are. The miracles of Soviet science. In this cellar, we separate streams of atoms. In that cellar, Mueller explodes farm animals.'

'Why?'

'Bombs produce pressure waves, as well as light and heat.

Different ones at different altitudes. Beyond a certain distance, the heat dies away and it's the blast wave that produces the lethality. So Mueller calculates it.'

'*Lethality?*'

Axelrod made a moue of distaste.

'Kill-zone radius. Casualty estimates, that kind of thing. The military men are interested in such matters. Nothing to do with atomic physics, of course. But that barometric sphere needs power, and space. So it was built in here, in the basement, next to our calutron. And so we live. Neighbors to our German destroyer of livestock.'

They walked to the lobby in silence. The linoleum gave way to a carpeted corridor that brought them once more to the turnstiles. Vasin spotted Kuznetsov slumped on a bench, his nose in a book. He stopped just out of his handler's earshot.

'What do you think happened to Petrov, Dr Axelrod?'

The scientist's voice, when he spoke, was trembling.

'It was not an accident. Petrov did not commit suicide.'

Axelrod turned to go, and Vasin recalled something Adamov had said about Petrov's final dinner. About the only guest Vasin had yet to meet, dead or alive. Colonel Pavel Korin.

'Wait. What's in Olenya?'

The scientist hesitated.

'I'm not sure I'm supposed to . . .'

'Come on. This isn't a test of your discretion. You failed that one about fifty times already.'

A look of alarm crossed Axelrod's pinched face.

'Forgive me, Comrade. What is Olenya?'

'It's a naval air base up on the Kola Peninsula on the White Sea. It's where the bombers take off when we conduct atmospheric bomb tests.'

'Do you know someone named Korin?'

'Everyone knows Pavel Korin. He works on weapons delivery. Brilliant nuts-and-bolts man. But why?'

'I believe he's a personal friend of the Adamovs. How well did he know Petrov?'

Suspicion closed down Axelrod's face like a rattling shutter.
'Ask him yourself.'
'That's precisely what I want to do, but he's in Olenya.'
'Testing dummies of RDS-220, I imagine. I have to go.'
Vasin caught Axelrod's arm and held it, pulling him close.
'How do I get to Olenya?'
'It's a busy time. There are transport planes going up there round the clock. Get a travel order from your superiors. Now let me go.'
'My superiors want me to sign off on their suicide theory and go home.' Vasin released Axelrod's sleeve. 'Is that what you want me to do?'
The scientist sagged.
'Keep me out of this.'
'I will try. How do I get myself on one of those planes?'
'They take couriers on practically every flight. Find some documents to deliver to Korin.'
'On whose authority?'
Axelrod thought for a moment.
'Adamov signed an Institute pass for you, right?'
Vasin nodded.
'In Arzamas, Adamov's authority is all you need.'

IV

'Interesting chat?'
Vasin looked up from the notebook in which he was writing up his meeting with Axelrod. Kuznetsov lay on the sofa, his book folded on his chest.
'Lots of chemistry. Or maybe it was physics. Didn't understand much of it, to be honest.'
Kuznetsov sat up on the sofa and began playing with a box of matches on the coffee table, flipping it end over end with his index finger.

'Listen, Vasin. You twisted their tails over at the *kontora* yesterday. Zaitsev was spitting nails all afternoon after your stunt. Getting some big Moscow pinecone to call Adamov and make him jump like a schoolboy.'

'The General strikes me as a zealous comrade who takes his duties very seriously.'

'Spare me your bullshit, I'll spare you mine.'

'We agree. Life's better without bullshit.'

'Zaitsev's an old-school butcher. You saw those scars on his right hand. He boasts they are burn marks he got from a red-hot pistol.'

Vasin made his face blank.

'Red-hot,' continued Kuznetsov, 'from firing hundreds of rounds into the backs of men's heads. They were evacuating a prison near Minsk. Early in the war, during the Fascist advance. There was no transport, so Zaitsev ordered every prisoner liquidated rather than see them fall into enemy hands. He decided it was his duty to do the job personally. That's Zaitsev.'

'And why are you telling me this?'

'Because he hates your guts, Vasin. You tell me why. I can guess most of it. He doesn't want complications. You have some ideas about Petrov's death. Maybe he's scared of who you work for. Who *do* you work for, by the way?'

'The Department of Special Cases of the Second Chief Directorate of the Committee for State Security.'

'Okay, so I don't need to know. All I'm saying is you got our boss all riled up. You need to walk carefully.'

'I'm just doing my job.'

'Which is to discover the cause of Petrov's death?'

'Except I'm getting the feeling that pretty much everyone here would rather not know.'

Kuznetsov shrugged.

'There are lots of things in the world it's better not to know.'

'Is that a message from Zaitsev?'

'Don't be paranoid, Vasin. It's a message from me.'

'Okay. Tell me why you don't want to know.'

Kuznetsov grimaced.

'Because of Zaitsev, yes. But not for the reasons you think. Because of what this place stands for, the people who work here, and what Zaitsev could do to them. All this that we are building in Arzamas? It's so fragile.'

'I don't understand.'

Kuznetsov snatched up his copy of *Science and Life* magazine from the table. On the cover was a picture of a new Sputnik satellite, shiny as a child's toy.

'Look here: this ball of steel, made by Soviet hands, about to go up into the cosmos and orbit the planet. Don't you see? After revolution and war and bloodshed, finally we've done it. We've beaten history. A new world is being born, right here in the Soviet Union. And the capitalists see it, and they hate it. They fear it. The future is being forged in front of their eyes, for every citizen of the world. The capitalists have to destroy us, or see themselves destroyed. That is why we need *them*. The Golden Brains.' Kuznetsov jabbed a stubby finger in the general direction of Kurchatov Square and Adamov's Citadel. 'In this space age, the bombs they make are our frontline soldiers.'

'What's that got to do with Petrov and your butcher Zaitsev?'

'You're not listening. Can we say the death of Fyodor Petrov was an accident?'

Vasin raised an eyebrow.

'You think it was no accident. A suicide, perhaps? Maybe. You think, maybe not. If not suicide, then it must be murder. And who are your suspects? Petrov's colleagues. His comrades. His fellow scientists. He wasn't hit over the head with a vodka bottle behind a bus stop, was he? Let's say it was someone who had access to some of the most radioactive material in the USSR, and knew exactly how to use it. Now we report all of this to Zaitsev and tell him to get cracking. If this is now a murder investigation, what kind of a culprit does a man like our Minsk jailhouse butcher look for?'

Vasin sat down heavily opposite Kuznetsov. He could see where this was heading.

'Zaitsev reverts to what he knows. He looks for enemies of the people. Ideological impurity. Antisocial habits. Unsanctioned reading of ideologically unsound books. Pederasty. Anti-Soviet literature.'

'And he would find them.'

'This is Arzamas. Of course he would. Look, here's a subversive book, right here.'

Kuznetsov picked up the paperback he had been reading and ruffled its pages in front of Vasin's face.

'This was printed by Russian exiles in New York last year. *Vozdushnie Puti*, Air Routes. Look who's in it. Osip Mandelstam, the greatest poet you've never heard of. Letters by Isaac Babel. *Poem Without a Hero* by Anna Akhmatova. Marina Tsvetaeva, not the lyrical Silver Age stuff you've read, but her late stuff. Bitter as quinine. Brilliant. Powerful. It's all brilliant. But every word of this book is counterrevolutionary propaganda according to Zaitsev and our own *kontora*.'

'Where did you get it?'

'I got it from the *kontora*, naturally. Haven't you ever been to the library in the Lubyanka? It's well stocked. *Newsweek* magazine. *Paris Match*. All the Soviet émigré publications. We have to know our enemy, you see? Or you could take a look on Dr Adamov's own bookshelves, next time you pay him a social call. I'm sure you'll find they're packed with banned literature.'

'Okay, so Arzamas is a greenhouse. Free spirits and eccentrics. But what makes it fragile?'

'I've lived among these people for three years now. And let me tell you, pretty much every one of them is a social deviant in one way or another. Allow Zaitsev to run amok in the Citadel, and he'd have confessions of anti-Soviet activity out of half the staff before you can say "purge".'

'The *kontora* doesn't do purges anymore.'

'And where did you read that, friend? Does the fact that something is on the front page of *Pravda* make it true? You're so sure that times have changed?'

'You told me yourself that the work they do is so valuable they can do whatever they like. They're untouchable.'

'As long as they *succeed*. What happens when they fail? What happens if RDS-220 turns out to be a dud? Do you have any idea how many millions of rubles go into this program? No – nor do I. But it's vast. Eats up the military budget. Eats up the security budget. You think old butchers like Zaitsev like to see all of this power in the hands of these degenerates and cloud dwellers? You think they like to hear General Secretary Khrushchev tearing down the great Stalin? There are plenty who think this entire show would be better run by some square-headed military engineers.'

'If Zaitsev hates all these misfits so much, why is he protecting them?'

'He's following orders. But imagine the Golden Brains mess up this test of Adamov's device. We lose the nuclear race to the Americans. Then, let's say word gets out that there's a murderer on the loose here in Arzamas. Then the muzzle comes off the hound.'

'Your boss is the hound.'

Kuznetsov dropped his voice and leaned forward.

'Give him an excuse and he'll rip the guts out of this place. There will be no more of this.' Kuznetsov flourished the book once again. 'No more of the likes of them. Or me. They brought me in from a cushy posting in the *kontora* station in Berlin. In those days you could still pop over to the American sector for an afternoon. It gave me a taste for all this dangerous literature, I suppose. Subversive tastes, Zaitsev would call them. Do me a favor and back off. Petrov was a tragic suicide.'

'And if it was murder?'

'You think there's just one unpunished murderer loose in this country?' Kuznetsov reached to his shoulder and tapped it with a pair of fingers, signaling invisible epaulets, the universal vernacular gesture for secret police. 'In the *kontora*?'

Vasin leaned back in his easy chair and ran a hand through his thinning hair.

'Thank you for your confidence, Kuznetsov.'

'You just seemed like a man who might understand. Clever face on you. Could be misleading, of course.'

They both grinned.

'I need to make a personal long-distance call. Not from a *kontora* phone.'

'Ah.'

'No, really. To my wife.'

'Try the post office, just off Lenin Square.'

'Is it open?'

'This city never sleeps, old man. I'll drive you if you like.'

'Mind if I walk?'

'And if I said I did?'

'I'd call in the big pinecones.'

V

An oppressive evening quiet had settled on Arzamas. Two empty trams rumbled toward each other from opposite sides of Lenin Square and stopped side by side. They stood, humming, with their doors open, as though gathering strength to continue their pointless journeys. Eventually the doors clacked shut and they moved off, their women drivers blank-faced as shop dummies. The entire town seemed to have gone to bed with a book.

Hard, cold loneliness welled up inside Vasin as he walked toward the town center. In Moscow, Nikita would be getting ready for bed. Vera would be on the phone, gossiping with her girlfriends. Or watching television. He pictured his home, his life, without him in it. Vasin urgently wanted to hear his son's voice, to impress himself on his mind. I still am your father. Don't forget me, my darling boy.

Arzamas's Central Post Office loomed into view. Rows of chandeliers glowed in the deserted hall. Two service windows were still open for telegrams and long-distance calls. Vasin felt in his pocket for a coin and tapped the metal counter with it. The telephone operator bustled out of a back room, a bundle of floral print and peroxide hair.

'Long-distance call. Moscow. Urgent.'

'Personal or service?'

'Personal.'

The operator slapped a long form in front of him.

'Fill this out.'

Vasin had seen foreign travel applications that were shorter. Dipping an old-fashioned steel pen into an inkwell provided by the State for citizens' convenience, he filled out his passport details. The name and address of the party he wished to call, the purpose of the call, his rank and address and his superior's name, the number of minutes he wanted – six – to be paid in advance. He blotted the form and handed it back. The operator's fingers danced on a large abacus.

'Eight rubles thirty kopecks. It'll be booth three.'

'How long will it take?'

'As long as it takes. Authorization.'

Vasin settled wearily onto a long oak bench polished smooth by citizens' backsides.

What had gone wrong? He did not blame Vera. She had just become what he always knew she was, if he was honest with himself. She wanted what every Soviet housewife wanted: an apartment, a family, a new television, friends, acceptance. He'd wanted all that, too, at the beginning. They had both been twenty. Marriage had seemed a preordained stage of unfolding adulthood, like receiving one's first passport or Communist Youth League ticket. 1947: The old world had been broken, the new one not yet born. Everyone who had called themselves young during the war had been transformed by the horror and hardship. The old-young soldiers seemed to envy Sasha's and Vera's youth, their innocence. And inexplicably resent them for it. At the wedding, a rowdy affair that involved dispensing free booze to their neighbors, the drunk snaggle-toothed old women toasted happiness, love, and peace in the world. Vasin wondered if they had actually known any of those things.

For Vasin and his young wife, marriage was an escape to a room of their own. But in this room they had to learn to live with each other, as well as learn who they were themselves. In the subsequent years Vasin had changed. Something deeper than just marital boredom and the strains of the job.

'This one wants to rearrange the world to suit him,' Vera's mother had said one afternoon at Vasin's in-laws' tiny dacha as he sat drunk on homemade moonshine. 'Our dreamer.'

He'd been having some kind of political argument with Vera's father. Anyone cowardly enough to allow himself to be taken prisoner by the Germans was an enemy of the people, had been the old brawler's point. Screw 'em. Spies. The old man refused to believe Vasin that the world was moving on. That there were new rules. Now the nation's battle was no longer for survival but for progress, for plenty. Apartments. Moon rockets. Washing machines. Shorter working hours. Cheaper vodka. At least they'd agreed on that.

The ringing phone sounded like a fire alarm in the echoing marble hall. Vasin hurried to the wooden booth and closed the glass doors behind him. He picked up the receiver and heard several female voices on the line. The call had already been connected.

'Go ahead, caller,' said the loudest of the voices. There was no click to signify that the operator had disconnected.

'Vera?'

'So how are you, my *darling*?'

He recognized the sliding intonation at once. She'd been drinking.

'It's him,' she said to someone beside her.

'Send him to sit on a dick,' said a shrill woman's voice from the background.

'Vera. I hope you are well. How is Nikita?'

'Normal. And you, dearest? How's the trip? They put you up in a nice hotel? Soft bed? Clean sheets?'

'It's fine.'

'And have you found yourself a nice little tart there, Sasha?'

'Come on, Vera. Stop it. This is a trunk line.'

'I see. Staying faithful to your tart in Moscow? Katya? Katya fucking Orlova . . .'

Vasin slammed the receiver down, wincing.

'For God's sake, Vera,' he said, as though she could still hear him. 'What are you doing?'

He picked up the phone again, but the line had gone dead.

Vasin stalked out of the hall into the moonlight. The woman at the counter watched him go with a smirk.

Would the silent listeners on the line bother filing a report? Had they heard? Vasin imagined fingers clattering out a transcript on a typewriter, the carriage jolting as it bounced into uppercase to capitalize the proper names per standard procedure. Zaitsev's beefy fingers holding the paper. His thick chuckle. And when he reached the last words, KATYA ORLOVA, a ripe curse. The name that could cut off Vasin's life as abruptly as a finger pressed onto the cradle of a telephone.

Vasin forced himself to breathe and start again at the beginning. A tired operator in the bowels of some KGB listening station, eager to get back to her tea and gossip. The last of the day's hundred domestic disputes cracking over the wires. Most likely she'd log the call and forget it. Add a sheet to the *kontora*'s ever-growing Himalaya of useless information.

The windy expanse of Lenin Square opened up before him, lit by weak municipal lamps and a yellow wash of light spilling from the sheer glass facade of the Kino-Teatr Moskva. He looked up at the new moon, half-hidden in drifting cloud.

On the roof of the cinema, something caught his eye. A face, illuminated by the streetlights. A young woman was standing on the high parapet, looking down. He recognized the fashionably cropped hair immediately.

VI

Vasin's boots clanged on the steps of the fire escape, the leather soles slipping on wet steel. By the time he had reached the top he was doubled over, panting. The roof was dark and flat. In the center, at the front, stood Maria Adamova, framed in a halo of light like a tiny diva on a vast stage.

She stood motionless, hands deep in the pockets of a vinyl raincoat. Her head hung down, as though she was asleep on her feet. When Vasin stumbled she made no sign of hearing.

He reached the edge, about five meters away from her, and peered down onto the square. The drop was sheer.

'Maria Vladimirovna?' He spoke softly. 'We met, with your husband. My name is Vasin.' Her eyes flicked open as though a switch had been thrown. She turned her face to him, then the rest of her body. She swayed forward drunkenly before abruptly righting herself.

'Go screw yourself, *Chekist*.'

Her voice was thick as mud.

Vasin advanced a step.

'Get lost or I'll jump.'

She stepped up onto the narrow wall that surrounded the roof. Balancing herself with her arms like a child, she lurched forward toward the void. With her arms out and coat flapping in the breeze, she looked like a ballet dancer about to take a bow. She stared out onto the square, mesmerized by the drop before her.

Vasin darted forward three, four, five paces, then seized the smooth plastic of her coat. Maria tumbled sideways, arms still outstretched. She fell with her back squarely on the parapet and attempted to wriggle her body over, into the empty air. Vasin held her coat tightly. She landed a single, despairing kick on the side of his head, then rolled, defeated, into his grasp.

Sparks sprayed across Vasin's vision as he watched his hat spin like a coin and tumble into space. He sank to his knees and slumped his full weight onto Maria Adamova's slight body. Panting together, they gulped down cold air.

'Maria Vladimirovna, you've been drinking.'

'Go fuck yourself.'

Vasin took Maria by the lapels of her coat and swung her into a sitting position. Her head rolled as though it was too heavy for her neck.

'You need an ambulance.'

'I said . . .' Masha shook her head as though willing blood and consciousness to flow into it. 'No bloody ambulance.'

'You need help.'

Masha's small hands closed around Vasin's wrists and unpeeled his grip on her lapels.

'Like hell I do. They'll send me to the nuthouse. You want me to spend the rest of my life as a vegetable?'

A memory, sharp as a flint, chipped across Vasin's mind. A girl, blond and frail, with a leer of pain smeared across her face. Klara. His younger sister, born two years after him, used to follow him about like a puppy. Her first seizure struck soon after their mother announced that their father would never come home from the war. His mother's flat adult voice, clinging to dignity, speaking hollow words of sacrifice and glory. It was to Vasin, not their mother, that Klara had run to soak up her tears. And then, a few days later, the fit came with terrifying violence, twisting her body and face into a grotesque, trembling arc. We must trust the doctors, Vasin's mother had said. Soviet science is the best in the world. They'd recommended electric shock therapy.

Vasin never saw Klara smile again. After a few months at the hands of Soviet psychiatrists all she could do was moan incoherently, like an overgrown infant. He turned her thin hand in his and saw marks on her wrists where they'd tied her to the bed frame. In his narrow bed at night, Vasin fantasized about rescuing her. He was sixteen – old enough perhaps to pass for a doctor if he wore a white coat and stethoscope. Also old enough to wonder where they would hide, how he would look after this wreck of a girl in her urine-stained nightgown. Then she stopped eating. Her fine blond hair fell out, and they put her in an isolation ward and wouldn't let Vasin visit anymore. 'The body has already been cremated,' Vasin's mother told him. Not 'Klara', but 'the body'. It was easier for them both to believe that Klara was no longer present in her tortured body while it underwent its final indignities.

Spend the rest of my life as a vegetable.

'Okay, Maria. I'll take you home.'

'How the fuck did you even get here? You been following me? You're not taking me anywhere.'

Maria made a defiant attempt to shiver some life into her leaden limbs. But her strength was spent and she slumped slowly to the ground. Scrabbling his heels along the wet roof, Vasin wriggled into a position by her side, his back braced on the parapet. She subsided into his lap with a shiver, balling her fists and tucking them under her chin like a toddler. In a moment, she was fast asleep. Vasin felt the wetness from her hair soaking into his trousers. He tugged one side of his mackintosh out from under her body and covered her as best he could.

Vasin lit a cigarette and smoked it with his left hand. He laid his right on her skinny, gently heaving shoulder. On the cushion of his lap, Maria began to snore.

Vasin chuckled to himself. Oh, Vera, he thought. Oh, Orlov. Oh, bloody Zaitsev and Adamov and all the rest of you. What, exactly, would you make of this?

The minutes passed. Young voices emerged from the Café Kino, their comradely goodbyes echoing around the empty square. Somewhere far away, thunder rolled over the forest. Maria's weight was pressed awkwardly on his right leg, which began to lose sensation. The breeze freshened. There were droplets of rain on the gusting wind, and from the darkness came the hiss of rain on rooftops.

'Okay, girl. Time to get you home.'

Stiffly, he untangled himself from the unconscious Masha and laid her on her side, her face on an outstretched arm. Kneeling beside her, he cradled her head in the palm of his hand and slid another under her armpit.

'Fedya?' Masha's eyes remained closed, her voice but a whisper. 'Fedyechka? Is that you?'

Vasin tried to pick her up, but her body was a dead weight.

'*Ya.*' Vasin murmured the monosyllable softly. 'It's me.'

'They said you had been poisoned. But here you are, thank God.'

'*Ya zdes,*' Vasin answered. 'I am here.'

Maria's face softened into a smile as she sank back into unconsciousness.

The drizzle pattered on her plastic coat and the cold rain rolled down her face like tears. Vasin shook her, but she did not stir.

He managed to drag her most of the way across the roof to the fire escape. Exhausted, he peered at the steel staircase, winding down six stories to the ground. He thought of heaving her over his shoulder in a fireman's lift, but the stairs were too steep and slippery. Vasin propped her upright against the steelwork of the staircase.

'Maria, I'm going to get help. I need help to get you home.'

Her eyes fluttered open.

'You!' she breathed.

'I'm going to call the police. I can't carry you.'

'Wait. *Wait.*'

Masha closed her eyes and rolled slowly onto her front with a curse. With a superhuman effort she struggled like a newly hatched butterfly to make each one of her limbs work in turn, first arms, then legs. She rose on her hands and knees and stayed in that position for a minute, head down, panting like a wet dog.

'Okay.'

Masha stood, very deliberately, and scraped the wet hair from her face.

'What are *you* waiting for?'

She turned unsteadily toward the stairs. By the time they had reached the bottom, her gait was rolling a little, but she stayed upright. She leaned against a wall, gasping.

Lenin Square was deserted. Vasin checked his watch. Half past twelve. A ten-minute walk to the Adamovs' apartment, and he would be rid of her.

A car rumbled into view. 'Shit,' Masha said and gripped Vasin's arm with two small, strong hands.

An old-model Volga police car entered the square from Engels Avenue, waddling on soft tires across the tram tracks.

'Here. Quick.'

Maria's hands darted to the sides of Vasin's face, closing behind his ears. She pulled his head down to hers. Vasin felt the heat of her lips on his as the yellow headlights illuminated them. Her face had a feral smell of sweat, as well as something sour and chemical. After a kiss that lasted seconds he pushed her away, staggering backward. The squad car had already swung past them and was cruising away down Lenin Prospekt.

'What was *that*?'

Masha had also recoiled and leaned against a young plane tree.

'Insurance, *Chekist*,' she said eventually.

'Excuse me?'

'The *musor* – the cops – saw us. Together. What are you gonna tell them when you turn me in?'

Vasin was speechless. He had been checkmated, perfectly, by an addled girl in a single move.

'So you can fuck off and leave me alone, Comrade *Major*.' Maria Adamova turned, gripping the narrow tree trunk, and vomited almost delicately into the flower bed.

'You can't keep me from following you, wherever you're going.'

Masha wiped her mouth on her plastic sleeve and looked Vasin up and down with undisguised distaste.

'So you're going to stick to me.'

'Like a tail to a vixen.'

Masha cracked a smile, despite herself.

'Come on, then. I know where we can get some coffee.'

Vasin took Maria's arm and steered her out across the buffeting wind that swept Lenin Square. His lost hat was scampering in the wind like a puppy.

VII

The long rows of wooden huts were unlit, and the gravel courtyard in which they stood was infused with the stink of engine oil. On the porch of the largest hut Masha fumbled for a light switch that illuminated a single, flyblown bulb.

'Keys are in the pot.'

She pointed to a flowerpot with old paintbrushes sticking out of it wedged on a windowsill. Vasin took hold of the brush handles and found that paint and brushes had long ago dried into a single plug. He picked it out and found the key below. The padlock snapped open with a well-oiled click.

Masha found a table lamp and switched it on. The wooden barrack was surprisingly spacious. The room was entirely lined with bookshelves, desks, and filing cabinets. To one side stood a draftsman's sloped sketching table, with a ruler and set-square attached to its surface with wires. In the far corner stood an Army-type truckle bed with the gray blanket neatly tucked in. A row of tall military boots stood in front of an old oak wardrobe.

'Who lives here?'

'A good man. He lets me use it when I need to.'

'Where has he gone?'

'Olenya. It's in the Arctic somewhere.'

'What's his name?'

Vasin immediately regretted the question. A fragile understanding had grown between them as they had walked through the nighttime streets. Now she snatched it away.

'Ask your colleagues, Comrade Chekist.'

She tossed her wet, ripped mackintosh onto the floor and collapsed on the bed, the fight drained out of her.

'Well,' she said. 'Coffee?'

The tiny kitchen area was as neat as a ship's galley. Vasin found a pair of tin mugs, a steel jar of Cuban coffee, condensed milk, a paraffin Primus stove. He shook it to check it was full and pumped up the stove's pressure with the little plug pump. Then he tipped a little alcohol from a corked bottle onto the burning cup and lit it to produce a fine blue streak of flame. A satisfying knack, remembered in Vasin's fingers from childhood. He opened up the vents of the main burner gently and blew them all into life, then filled a dented pot from the tap and set it on the roaring stove.

'We all used to live like this, you know.'

Maria's voice sounded unnaturally loud in the dark hut.

'Adamov. All the old scientists. The engineers. When they first started working here after the war, everyone lived in barracks like this one while they were putting up all those new buildings. There used to be partitions in them, all the way down. Tiny, little rooms. One bathroom, down there at the end. One kitchen.'

'Even you?'

'Yes, even us. They offered Adamov a proper apartment in the old town. He said, not until everyone has one.'

'Very admirable.'

The water began rolling in the pot. Vasin stirred in four spoonfuls of coffee and turned down the Primus.

'It was luxury,' Masha continued, her voice petering out as if she was talking to herself. 'Compared to Leningrad, everything seemed luxurious. Adamov found me there, you know. After he came back from the North.'

The North? Vasin popped a triangular hole into the lip of the condensed milk tin with a can opener and dribbled the white syrup into their coffee mugs.

'And your anonymous friend who still lives here?'

'He stayed on in the barracks after everyone else moved out. Took over the whole hut and burned the old partitions in his stove. He preferred honest wooden planks to concrete walls. Wise man. He knows everything that goes on around here.'

'They didn't make him move?'

Masha gave a snort.

'Haven't you worked it out yet? The Golden Brains of Arzamas do exactly as they please. They want to live in an old wooden barrack full of cockroaches? Of course, sir. Crazy? Sure. But we are all very, very crazy.'

Abruptly Masha flipped herself over in bed and rose on her elbows, her face in the light.

'Crazy-crazy-crazy,' she hissed, widening her eyes in an exaggerated mime, watching Vasin in the blue light of the Primus flame. 'Fucking cray-zee.'

He brought the hot mug to Masha and squatted next to the bed, his face level with hers. The mask of mockery fell away. She accepted the steaming coffee and took a sip.

'Damn, that's good.'

Vasin sipped too. Army-style: strong, milky, and sweet.

'You were happy here.'

'Yeah. Happy.'

'So, how did you end up on the roof of the Kino tonight?'

Her eyes met Vasin's, and he saw himself, tiny, in her pupils, suspended like a prehistoric fly in amber.

'Ask me some other time.' She swigged down the last of the coffee. 'You'll hear the vixen's tale.'

Vasin straightened and drained his mug of coffee. Masha lay huddled in the narrow cot, her eyes trembling in apparent sleep. He turned to leave, but habit made him tug the wardrobe door. It opened stiffly to reveal the dress uniform of an Air Force colonel, with an engineer's crossed hammers on the breast. He slipped a hand inside and turned the lining of the breast pocket out into the light. The owner's name was written in the military tailor's square hand. KORIN P. A.

The freezing night air stung Vasin's bruised eye, but he was in no hurry to return to Kuznetsov's. He lit an Orbita and puffed it, enjoying the silence and solitude. He was just about to start for home when he heard it: a thin whine, almost like the buzz of a mosquito. The noise was so faint that he was ready to dismiss it as a ringing in his ears when it changed pitch. Higher, lower, and higher again. It was coming from somewhere inside the hut.

Ducking out of sight of the windows, Vasin followed the sound to the back of the cabin, near the kitchen area. Through a frosted pane he saw Masha, wrapped in a blanket, squatting by an open cupboard. Her hair was haloed by faint electric light. Abruptly, the whine resolved into a man's crackling voice. Vasin strained to hear, but caught only one phrase that came directly after a snatch of music.

'This is the Voice of America . . .'

CHAPTER FOUR

TUESDAY, 24 OCTOBER 1961

SIX DAYS BEFORE THE TEST

I

Vasin's portable alarm clock rang under his pillow at three in the morning. He crept into the bathroom and closed the door carefully before switching on the light. In front of the mirror he prodded a ripe blue bruise that had hatched on his temple where Maria Adamova had kicked him. Nothing to be done about that. Apart from lie.

Vasin could feel her strong fingers on the side of his head, the urgent gesture with which she had pulled his face to hers. He thought of her and Adamov sitting at their grand dining table, still as waxworks. The slow swing of her walk as she had led him out of the apartment. And later, the fury in her eyes. The damage. His sister's doomed, defiant spark.

'Fedya?' she'd called him on the roof of the Kino. 'They said you had been poisoned.'

Fedya. The diminutive of Fyodor. Fyodor . . . *Petrov*?

Vasin examined his own face, bruised and puffy with sleep, in the mirror. Who was there to feel sorry for him, now that Vera had unleashed her hatred? Only his son, Nikita. Once a week they would escape for a few hours into the city. Gorky Park was a favorite, watching their fellow Muscovites stroll along the wide promenades. Ever since the boy had conquered his fear of the squeaking Ferris wheel many summers before, they rode it ritually on every visit. Now that he was thirteen, the ritual seemed defunct, but when

Vasin suggested doing something different, Nikita just shrugged, and they ended up on the wheel anyway. What had he done this past Sunday instead? Something miserable, doubtless. Piano practice. Visiting his grandmother.

Vasin could be back in Moscow in time for next Sunday's jaunt with Nikita if he just did what General Zaitsev demanded.

In his mind's eye, Vasin scrolled through his colleagues, old and new, putting each one opposite him. What would you do in my shoes, Comrade? And you? The lumbering old alcoholics of the Moscow homicide squad, the smooth-faced new KGB men in their apparatchik suits and smelling of Troinoi eau de cologne. They all had the same answer. All of them would do exactly what they did every day. Sigh. Shrug. And sign. Of course, they would all sign the report.

Perhaps a year ago, new in the State Security job, Vasin would have done the same. His natural curiosity, a compulsion to follow the evidence to its logical conclusion combined with his urge to look around the next corner, was one thing. But until this moment that hunter's instinct had always been rooted in good sense. His family needed to be looked after. His precious career needed nurturing. Vasin had watched the brown folders disappear inside General Orlov's safe in complicit silence. 'The higher morality of the Party,' Orlov had called it. The Air Force General's young wife allegedly murdered at her dacha by armed intruders, although the nine grams of lead in her brain had come from her husband's service weapon. The Politburo member's heroin-addicted daughter found comatose in the studio of a well-known subversive artist, the heroin supplied by the nephew of a Party bigwig in the Uzbek Soviet Socialist Republic. The sins of the Soviet Union's officially nonexistent ruling class that Vasin had uncovered, consigned to Orlov's steel-lined chamber of secrets, as explosive as any bomb ever made at Arzamas.

But now the anchors of Vasin's world had been abruptly tugged loose. Vera had threatened divorce. She probably meant it. She would get to keep the apartment, and Nikita. And how long could it be before General Orlov found out about Vasin's affair with his terrifying Valkyrie of a wife?

Vasin's life had become a barge drifting into a quickening current of chaos. After Orlov discovered he had been cuckolded, Vasin could be sure that his next posting would be as director of a penal labor colony up in Magadan. Or no – Orlov, with his priestly theories of justice and retribution, would take some time to consider Vasin's punishment and make sure it was appropriately biblical. Back it up with a trumped-up corruption charge that would give him no choice but to accept. Something resonant with Arzamas, perhaps? Commandant of a uranium mine on the upper reaches of the Kolyma River. He'd be a bald, poisoned wreck in five years.

A pulse surged through Vasin's bruised temple, sending a bloom of pain through his face. Perhaps he would be able to unearth something in Arzamas. Something to trade with Orlov. With the *kontora*. With anyone powerful enough to save his doomed carcass from Siberia.

Vasin had one more day in Arzamas.

It was time to get up to Olenya and talk to Pavel Korin.

He left Kuznetsov a scribbled note:

Gone sightseeing. Back tonight. Don't wait up.

II

The airstrip at Arzamas was a simple affair. A runway and a row of hangars, a prefabricated concrete terminal and control tower. A pair of Antonov-8 transports stood with their rear ramps down on the apron, spilling bright light onto a team of loaders manhandling cargo.

The spotty-faced duty sergeant scanned Vasin's KGB identity card, his Institute pass signed by Adamov, and his travel order to Arzamas without curiosity.

'Got a billeting order, Comrade Major? It's jam-packed in Olenya these days.'

'I'm not staying. In and out. Classified documents from the Lubyanka for Colonel Korin.'

The kid nodded reverently and copied Vasin's details into a flight manifest. Vasin took the stamped piece of cardboard labeled 'Movement Order' and tried to make himself comfortable on a hard bench in the waiting room. In the pocket of his uniform greatcoat a paper bag of boiled sausages jammed into bread rolls from the station café warmed his thigh.

'Passengers for Olenya!'

Vasin woke abruptly and looked about the waiting room. As he had dozed, a dozen fellow travelers had gathered about him. They were a motley group of civilians and military engineers, some clutching document cases and blueprint rolls, others nursing Army kit bags. Nobody spoke. Vasin's breakfast had gone cold in his pocket. They filed out onto the tarmac into a swirl of fine, wet snow, guided by the yellow glow of the Antonov's lights. There was still no hint of dawn in the blackness of the eastern sky.

The steel-ribbed interior of the aircraft resembled a ship's hull. It was stacked to the ceiling with wooden crates secured by webbing and straps. Canvas seats folded down from the walls. Vasin squeezed in beside an older man who wore Arctic gear of padded breeches, full-length sheepskin coat, and Army-issue fur hat. Vasin's neighbor looked him over wordlessly as he stuffed grimy wax plugs into his ears. The ramp creaked shut with a squeal of hydraulics, and the cabin lights went off without warning. The transport's engines rose to a roar, and Vasin could see the snow streaming into a roiling tunnel behind the propellers.

It was Vasin's first time in an airplane. He felt anxiety tighten his bladder as the Antonov lurched into the air, climbing steeply into the gathering snowstorm. He had left the earth and was now in the vertiginous, mechanical world of the cloud dwellers. The high places from which shiny planes dropped deadly bombs on mere earth dwellers below. Involuntarily Vasin grabbed his neighbor's arm. The man smiled, steel teeth gleaming in the cabin's red emergency light. The plane steadied as they rose above the cloud canopy. The moonlight illuminated a vast prairie of cloud. It was pale silver, thought Vasin, like the landscape of a dream.

III

The Arctic sun dawned over the lake at Olenya, a pale streak in the southern sky. The men who looked up from their work saw it as an elongated, almost rectangular patch of light shimmering above the horizon. The meteorologists called it a 'polar mirage'. Not the sun but a mere reflection of it rising on some distant, slightly less frigid corner of the Soviet Union far to the south. A false dawn, then.

Nature mocks mankind, thought Vasin as he crunched across the snow toward a line of refueling trucks by Olenya's runway. It mocks man with visions of what he craves the most. In the desert, mirages of pools of cool water. Here, in the Arctic, men see visions of the life-giving sun peeping impossibly around the curve of the earth.

In a parked bus a row of pale, youthful faces pressed up against the windows like anxious young mothers seeing their babies off to school on the first of September. Engineers, Vasin guessed. He pulled open the door.

'Colonel Korin?'

'He's out on the apron.' A slight kid, swaddled in Army-issue sheepskin, finished stuffing papers into a leather document case. 'I'm going over there now.'

They walked side by side across hard-packed ice. On the runway, floodlit by a circle of arc lamps creating a steaming halo in the night's gloom, stood the pencil-like form of a Tupolev-95 bomber. The aircraft had been painted a brilliant metallic white all over. But the bomb bay doors and part of the fuselage of the once-sleek aircraft had been cut away. The payload she carried stuck out underneath like a pregnant belly: it was unmistakably a bomb, almost cartoonishly so, with a fat belly, snub nose, and fins. The steel casing was at least eight meters long and two wide.

'Lord,' said Vasin.

The young engineer grinned.

'Big, isn't she? We spent all night loading her. Twenty-seven tons of cement inside that baby. Nearly two and a half times the bomber's normal weapon load. More than triple the size of any device we've ever loaded.'

Cement? Of course. Dummy bombs, in preparation for the test.

'That's Korin.' The kid pointed to a powerfully built man with a thick gray beard who stood framed in a blaze of arc light. He wore uniform breeches under a heavy sheepskin aviator's coat. His angular face had once been handsome but was now hard and dented as an old cast-iron stove.

A huddle of soldiers and sergeants squatted by a radio truck muffled in their winter coats. Vasin joined them. Voices came crackling over the radio.

'Final request for visual confirmation from ground crew.'

Korin strode over and leaned into the truck, taking the radio receiver in a gloved hand.

'Colonel Korin confirming all normal. *Nasha detka krepko pristegnuta.* Our kid is tightly strapped in.' The pilot's voice came loud and steady over the radio's hiss.

'Load trim checks complete. Starting engines.'

One by one, the pilot fired up the four turboprop engines, each of them powering two oppositional propellers. Korin pointed both his thumbs up to the sky and the pilot did the same. Vasin had no idea what the gesture meant, other than it was something Yankee. The overloaded aircraft began its slow taxi toward the runway.

Something like a real dawn had come by now. The rising sun illuminated the steaming breath of men and the exhaust boiling from the aircraft's engines in a weak golden light. The men around

Vasin watched the Tupolev trundle down the extended runway with professional eyes.

'*Shas' yebnit*,' grunted one of the old hands through missing front teeth. 'She'll fucking crash. It's too damn big.'

'*Molchat'! Yazyk t'e vyrvu, pizdyuk yebuchy.* Shut up or I'll rip your tongue out, cunt!' called Korin. Vasin recognized the argot of the Gulag immediately.

The pilot lined his aircraft up on the runway and braked for the final preflight check. Korin raised his field glasses and watched the engines winding up to full power for takeoff. Maybe she *is* too damn big, Vasin thought, watching the belly of the bomb accelerating barely two meters clear of the tarmac. But no. The Tupolev lifted off late, but gracefully. In a slow arc of black exhaust she turned north, following the receding night.

'Colonel Pavel Korin?'

'Who are you?' The voice was deep and imperious.

'Major Alexander Vasin.'

'What do you want?'

'I want to talk to you about Maria Adamova.'

Korin made no perceptible movement, but Vasin saw concern pass across his face. The Colonel's eyes flicked to the green KGB flashes on Vasin's uniform greatcoat.

'Has something happened to her?'

'No. She's okay. But it almost did.'

'Who are you, Major Alexander Vasin?'

'State Security.'

The Colonel made no attempt to disguise the distaste that pulled at the side of his mouth like a fishhook.

'Don't fuck with me, Major. What's wrong?'

'Do you want to talk about it here?'

Korin looked around. Black exhaust smoke was drifting across the airfield. Fuel trucks were revving up one by one to return to base. He checked his large aviator's watch.

'The *kukla* – the dummy – drops in thirty-one minutes. We can talk after that. Come to my quarters after breakfast.'

IV

Korin's billet in Olenya was an anonymous wooden building just like his digs in Arzamas, knocked together and painted a dull military green. A trickle of smoke came from the stovepipe and frost had spread in fantastic patterns over the window glass. Vasin knocked loudly on the door. Silence. After her brilliant dawn entrance, the sun had ascended into a bank of unbroken cloud, leaving Olenya to make do with a pale wash of gray light.

A polished black limousine turned in to the yard. The car was ungainly as a truck, but its chunky lines had an old-fashioned elegance. The ZiS – the largest sedan ever produced by the Stalin Factory – was the preferred conveyance of people's commissars of the prewar generation. The car rolled to a halt in front of the hut, and Korin climbed out of the backseat. He gestured to Vasin to come inside and stamped up the steps after him. Small puffs of sawdust floated down from the ceiling as the front door slammed behind them. Korin shrugged out of his heavy coat, hauled off his boots, and turned to Vasin, fists on his hips, formidable even in his stockinged feet.

'Okay. Now tell me.'

'Maria tried to kill herself,' Vasin said. 'Last night.'

Korin remained impassive.

'How do you know?'

'I was there on the roof of the Kino Moskva. I talked to her and brought her down. She asked me to bring her to your place, and not to report it.'

'And you didn't?'

'No.'

'Wasn't it your duty?'

'Yes. But I felt sorry for her.'

'*Sorry.*' Korin's voice was flat with disbelief. 'I'm sure you will put this experience you had with Comrade Adamova to its proper use.'

Vasin knew better than to bother protesting his good intentions to a hard old bastard like Korin.

'Spit it out, man. Why are you really here?'

'You saw Fyodor Petrov at the Adamovs' the night he died.'

'I did.' Korin's voice betrayed nothing.

'Just before he went home and committed suicide.'

'If you say so.' Korin gave a contemptuous grunt. 'Major, are you planning to take me in for questioning?'

'No.'

'Then that will be all. I have to be back at the workshop.'

Vasin turned his cap in his hands. He'd encountered men of Korin's type before. A survivor of the NKVD's torture cellars, if he had to guess. Unbending as an old oak.

'You were in the camps, Colonel.'

'What's it to you?'

'Kolyma? Magadan?'

'Does it make any difference?'

'Would you believe me if I said that those times are past? I was a cop until a couple of years ago. A police detective, not KGB.'

Korin remained silent.

'I don't think Fyodor Petrov killed himself.'

Again, silence.

'Was he close to Adamova?'

'Leave Maria alone.'

'I believe that you and she are good friends. She told me she admires you. So I'm asking you if she was involved with Petrov.'

Korin ran a stubby hand through his beard.

'If I tell you, will you promise to stay away from her?'

'I'm not here to lie to you, Colonel. I can't in conscience promise that.'

To Vasin's surprise, Korin grinned, and slipped into the familiar form of address.

'*Conscience*, eh? You wouldn't have lasted long in 'thirty-seven, little pigeon. You "can't in conscience"! Ha! That isn't the way you Chekists used to talk.'

'Times change. Even the *kontora* changes.'

'Listen to me then, Major, if times are so different. I'll tell you why you should stay away from Masha, and we'll leave it to your *conscience* to decide. Okay?'

Korin made his way stiffly to the kitchen area at the back of the hut and rummaged for a tin of tea. He pointed to a Primus stove, even more battered than the one in Arzamas. Vasin noticed that three fingers of his left hand were missing their tips.

'Light that.'

He tossed Vasin a box of matches and held up his damaged hand.

'Frostbite. Fiddly thing, that stove.'

Korin brewed his tea strong and black. Vasin noticed that he held his mug on the insides of his cupped hands. An old convict habit born in freezing places where every gram of heat is precious.

They sat on a pair of sagging armchairs in front of an incongruously splendid new radio.

'Masha lost her family in the war. I lost my family in the war. But she was a smart one. Just after the war I agreed to teach an evening class for an old professor of mine who was sick. Foundation course in applied mathematics at the Engineering Institute in Leningrad. Masha was the bright button of the class. Good heart too. About the same time, Adamov was looking for people to come to Arzamas to build the first generation of . . . devices. I agreed to come and introduced the two of them after a lecture. She caught his eye. Long story short, they got married. We all ended up in Arzamas. So now we're family, Masha and me. Did you know that she survived in Leningrad through the German blockade?'

Vasin briefly considered trying to bluff, but instead shook his head.

'During the heroic siege, the whole city was cut off by the Fascists for nine hundred days. Millions of people were trapped without food. Masha's parents starved to death. Left her alone to fend for herself. Do you know how people survived? You don't, because it's not in any history you would have read. They ate corpses, boy. By the time the authorities rounded up all the children at the end of the blockade, they were wild as rats, and twice as hungry. When I first met Masha, she looked like a scarecrow. Pair of green eyes on a sack of bones. Her stammer was so bad she could barely speak. I'd give her chocolates and she'd turn away and wolf them down. In the classroom she was quiet and attentive. But outside she would get into fights. When Adamov brought her to Arzamas, she didn't trust anyone. Tough as you like, in many ways. But her mind is . . . broken. Even after all these years, she's still on the verge of a collapse. That's why you need to leave her alone.'

Korin continued before Vasin had a chance to speak.

'If what you say is true, Major, you did a good thing when you brought her to my place instead of handing her over. I'm telling you this so that you understand. Because if those headshrinkers and trick cyclists ever get hold of her, they'll lock her up and electrocute her brain. Make her one of the living dead. The old man wouldn't be able to bear it.'

'The old man?'

'Adamov needs Masha. And we need Adamov. The whole world needs Adamov.'

Korin's words hung in the air.

'The world?'

'Young man, either you're stupid or you're playing a game that's too clever for me. *Why?* Because of RDS-220. The Big Bomb, the kids call it.'

'How is it different from all the other bombs?'

The old man wiped a hand across his face.

'Where did you say you'd come from?'

'I didn't. Moscow sent me. Special Cases.'

'I have a special case for you. Stoke up that stove. The wood's outside, under the porch.'

The split pine logs were full of sap, and left white stains on the wool of Vasin's greatcoat. A group of soldiers slouched past, smirking, as Vasin stumbled up the steps, dropping logs as he went.

'Where did you spend the war, little pigeon?'

'At school in Moscow. My father was a military engineer.'

'One of us! Where did he study?'

'Moscow Higher School of Aviation Construction. On Leningradsky Prospekt.'

'Ha. In the thirties? I might well have taught him.'

Korin took out a *papiros*, the worker's cigarette that consisted of a long cardboard tube with five centimeters of powerful dark tobacco at the end. The brand, Belomorkanal, was named for Stalin's great canal from the White Sea to the Baltic, built by convict labor. Punishment put to the good of the Motherland. Vasin had met some old convicts who'd worked on it. Korin puffed the *papiros* into life, moving it from one side of his mouth to the other as though he were chewing a straw.

'I was sent to Murmansk in 'forty-two. Assembling American Douglas bombers after they arrived from Scotland on the Arctic convoys. Getting them into the air. Training our pilots to fly them, our mechanics to arm them, adapting them to carry our heavy munitions. Not a bad job, while it lasted. But one day I flew a mission running air support for a British convoy. The pilot was an American kid, younger than me. He was teaching us to use the bomb-sighting apparatus to spot U-boats and drop depth charges. You've got to fly in low, surprise 'em and catch 'em before they dive. But we found a German destroyer instead, and it raked us with antiaircraft fire. Dan Bilewsky was the Yankee pilot's name. His people were Poles, I suppose. Anyway, we were a few miles south of Tromsø with the starboard engine on fire. Bilewsky managed to gain enough height to get us back to the coast. Best piece of flying I've ever seen. Corkscrewed up on one engine. As soon as we were over land, he ordered us all to bail out. We were low – only about two hundred meters – but we jumped. I was the last out before the engine detached and put the crate into a left-hand spin. Uncontrollable. Threw me upward on the roll. I thought, Well, that's your luck, Korin. You jump out of a fucking plane and fly up instead of down. Anyway, my parachute just had enough time to open, and I landed in a snowdrift. Walked away with bruises.'

'And the pilot?'

'Killed, of course. Don't interrupt, listen. My point is from that day onward, I saw war. The Norwegians took us prisoner. Me, the two bombardiers, gunner, navigator. Handed us over to the Germans. Our navigator had broken his collarbone, so an SS sergeant shot him in the face. Casual-like, as if he had other things on his mind. Then they put us on a stinking freighter to Riga, where they needed slave labor. We were with a hundred other Soviet prisoners. None of us had any food for a week. When we arrived, they put us to work digging mass graves. Burying the corpses of partisans and Jews. Cold. Hunger. You don't know what those words mean. But I do.

'Anyway. By the next fall our armies were pushing down from Leningrad. Our aviation blasted the shit out of the port of Riga for three weeks straight. You could get warm just by standing in the cindery wind that blew from the city. The Fascists were panicking, scrambling to save their backsides. One morning we woke up in the cattle shed where we'd been billeted and discovered our guards had gone. Just like that. They couldn't even spare the bullets to shoot us.

'Our guys were exhausted and starving, ready to die where they lay. I took up with a group of Yakuts. Tough bastards, these Siberian hunters. They robbed our sick of whatever food they could find hidden in their filthy clothes and we struck out to find our lines. I tell you, when I saw that first red star on a Soviet fur cap I cried like a baby. Malyshev's boys, the Fourth Shock Army. The officer was from Odessa, and he gave us some pork fat they'd captured and propped us in a row by a fire. We leaned against each other like frozen laundry. I ate like a wolf, then puked it all up again. It was the happiest hour of my life. Until you sons of bitches came. The Chekists arrested us for desertion. But that's a different story. The point is: I know what war is.'

'And now you make bombs.'

Korin spat out a piece of tobacco and leaned forward so that his face was centimeters from Vasin's.

'We make bombs so that there won't be another war.'

The words came with such vehemence that Vasin felt Korin's spittle landing warm on his face.

'RDS-220 will be the biggest thermonuclear bomb ever built. Three thousand times more powerful than the Yankees' Hiroshima bomb. At least. Maybe five thousand. A hundred megaton yield. That's what Khrushchev ordered up. That means it has the explosive force of a hundred million tons of high explosive. Twenty times bigger than any atom bomb that's ever been detonated before. A bomb that can kill every living thing in a two-hundred-kilometer radius.'

Vasin struggled to take in the numbers.

'And why are we making this infernal machine? Because it will be the end of war. You only fight if you think you can win. RDS-220 transforms war into suicide. There can be no victory, only total annihilation. Do you see? We are nearly there. Peace, forever. This is why the world needs Adamov.'

The old man swigged the dregs of his tea.

'What has this to do with Maria Adamova and Fyodor Petrov, Colonel?'

'Maria is a good and faithful wife. That's all you need to know. It's also the truth.'

Korin put his empty mug down on the table with an emphatic clunk. The conversation was over. Obediently, Vasin stood.

'Just one thing more, Colonel. Was Adamov in the camps as well?'

'Of course he fucking was.'

'What for?'

Korin jerked to his feet, looming over Vasin and sending his tin mug rolling across the floor.

'Same reason any of us was in the camps. Some ambitious little fucker denounced him. Some *Chekist* had a quota to fill. That's what *for*.'

The old man's eyes were blazing.

Vasin bowed his head. 'Thank you, Colonel.'

'You leave Masha alone.'

'There may be a killer in Arzamas. You don't think it's important to find him?'

'Did nothing I told you sink in, boy? The only thing that matters in Arzamas is the bomb. The bomb that will end war. Forever. Can you think of anything more important than that?'

'I got it, Colonel.'

'He got it.'

Korin muttered the last words to himself. Exhausted, the old man subsided back into his chair, closed his eyes, and in a second was asleep. A snore rumbled inside his chest like a growl. Vasin loaded the stove with more logs, pushed the damper half-closed with his toe, and left the Colonel snoring in the warming room.

Outside, the pale Arctic sun appeared for a moment high above the surrounding buildings. But it gave no warmth.

PART TWO

BURNED AND BLINDED

We stopped at a checkpoint where we were issued dustproof jumpsuits and dosimeters. We drove in open cars past the buildings destroyed by the blast, braking to a stop beside an eagle whose wings had been badly singed. It was trying to fly, but it couldn't get off the ground. One of the officers killed the eagle with a well-aimed kick, putting it out of its misery. I have been told that thousands of birds are destroyed during every test; they take wing at the flash, but then fall to earth, burned and blinded.

ANDREI SAKHAROV,
Memoirs

CHAPTER FIVE

TUESDAY, 24 OCTOBER 1961, EVENING

SIX DAYS BEFORE THE TEST

I

As the Antonov-8 dropped beneath the low cloud cover, Vasin saw that Arzamas had been transformed by the day's snowfall. The airfield was a white scar across the black of the forest. Beyond, the city was visible as a lonely island of light in an ocean of woodland. The pilot fought crosswinds all the way down, landing in a jolting skid that had Vasin grasping the canvas of his seat. As the rear doors opened, a chill blast of wind sent pain throbbing across his bruised face.

To his relief, Vasin found no reception committee waiting for him. Zaitsev and his boy Efremov would doubtless be scouring the town for him but clearly hadn't yet checked the airport's movement orders. Good. He might be able to steal a couple more hours of freedom.

Vasin hitched a ride into town with a group of fellow passengers. After the numbing cold of the aircraft, the fierce heat of the little bus stupefied the group into the companionable silence of bathers in a hot sauna.

Vasin disembarked at Lenin Square. Evening was already closing over the city, and the trams were crammed with citizens hurrying home. The high windows of the Adamovs' apartment on Marx Street were dark. Vasin waited at the flimsy shelter of a bus stop at the end of the block. She was easy to spot, her stride as upright and self-conscious as that of a schoolgirl stepping up to the podium to collect a prize. She was dressed formally in a smart topcoat and head scarf, and under her arm she carried a fresh loaf wrapped in paper. She walked right past him, her gaze fixed on some point in the middle distance, not on the ground in front of her like the other pedestrians. Vasin caught up with her as she was about to turn in to the apartment building

'You look well, Maria Vladimirovna.'

She seemed unsurprised to see him.

'You look like someone kicked the shit out of you.'

'Someone did.'

Maria pursed her mouth and glanced up and down the street.

'Better come inside. You need to clean that up.'

They mounted the steps in silence. The apartment was dark and silent.

'Professor Adamov?'

She answered with an eye roll and pushed the door shut with a backward kick.

'At work.'

Snapping on the lights as she went, she led Vasin through the dining room and into a large, white-tiled bathroom.

'Sit.'

Vasin obediently perched on the edge of the cast-iron tub. Maria opened a large medicine cabinet and began rummaging inside. The shelves were crammed with medicine bottles, some with the plain typed labels Vasin recognized as being from the Kremlin Polyclinic, others in their original American bottles with the bright logos of Bayer and Merck. He peered over her thin shoulder at the labels: phenobarbital, phenothiazine, chlorpromazine. Tranquilizers.

Maria picked out a bottle of iodine. As the cabinet door clicked closed, her face appeared in the mirror, and for a long moment she paused to look at herself. Vasin, out of her eyeline, could not stop himself from staring at her reflection. Fine mousy hair, cropped short like that of a French tomboy film star. Thin, sharp little features. High cheekbones supporting those huge eyes, too big for her head. And her head too big for a thin neck and narrow shoulders, like a child's. Seeing Vasin watching her, she grimaced briefly. Tipping iodine onto a wad of cotton wool, she set to swabbing Vasin's temple with brusque, businesslike movements. The spirit burned hot on the bruise.

'You didn't report me?'

'I did not.'

'And now you've come to present the bill. Let me guess.'

'No, there isn't going to be a bill. I know what the nuthouse is like.'

'Sure, you are a very knowledgeable man. That is obvious to everyone.'

She recorked the iodine bottle with a sharp slap and turned back to replace it in the medicine cabinet. Next, Vasin knew, she would ask him to leave. He felt a sudden, piercing urge to return to the complicity they had shared in Korin's barrack.

'My sister died in an asylum. They tied her down and electrocuted her.'

Masha froze in mid-movement, her back still to Vasin.

'I haven't told many people that.'

'So why did you tell me?'

'I don't know why,' Vasin said. 'I wanted you to know I understand.'

Maria nodded. She seemed to be turning his words over in her mind, like a magpie examining a bauble. Vasin felt suddenly stupid, exposed. This is work, fool, he told himself. A wash of pain from his temple mercifully interrupted that train of thought. It had been a mistake to come here straight after Olenya. The conversation with Maria had wriggled suddenly from his grasp, like a stick in a fairy tale that turns into a serpent. He stood.

'Maria Vladimirovna. May we continue our talk later?'

'Wait.'

'Your husband will be home soon.'

Maria's hand was on his arm, her grip tightening as he tried to pull free.

'Little rat.'

He looked away from her face, suddenly too close to his as she pulled herself toward him, her words hissing in his ear.

'That's what my schoolmates called me. Little Rat. And Stock Bones, for the packets of bones they sell for soup at the meat shop. Only twelve kopecks a kilo. Around here, my neighbors feed it to their pet dogs and buy chicken hearts for their fucking cats. But that's what I am, a bag of bones. A little rat.'

Her grip eased, and Vasin sat back down heavily on the edge of the bath.

'I see what you think. The Director's young, well-dressed wife, she gets whatever she wants. She lives in this huge apartment, spoiled as a butcher's cat. She's got no real problems, so she invents some of her own. I saw you eyeing those pill bottles. A hysteric. A junkie.'

'I don't see why you were on the roof. There is no obvious road from here' – Vasin glanced about the spotless bathroom – 'to the Kino. The reason is hidden, to me. As are you.'

Anger kindled in her pinched face.

'You think this is me? Think I want any of this? I do it for him. I wear these beautiful, foreign clothes from some commission shop on Oktyabrskaya for *him*. For Adamov. All of these vanities are his way of saying he will protect me from the world. And he tells the world, "This woman has power. Respect her." You have your uniform with stars on your collar. My husband has his own stars on his chest. I have my French clothes. You have no idea how much the other wives hate me. I can feel their glances on me, flung like spit. But I never show weakness. I learned that a long time ago. The pack always turns on the weakest, and then they die.'

'You're the weakest?'

'Obviously not. Otherwise I wouldn't have survived.'

'Survived what? Leningrad?'

Masha stiffened.

'You've been spying on me?'

'Your friend Pavel Korin is very concerned about you.'

'Where on earth did you find him?'

'Where you told me I would find him. When we had coffee at Korin's barrack.'

Masha folded her arms tightly across her chest.

'Christ. You told him that I'd brought a Chekist to his house?'

'Not sure I mentioned that part.'

Masha curled her lip in frank disbelief.

'I'm sure the next time you gentlemen get together to chat about the welfare of crazy Masha, you'll get around to it.'

'We weren't chatting about your welfare. We were talking about a death. Which is why I am here in Arzamas. Maybe you forgot. Perhaps you have things on your mind. Perhaps Fyodor was the thing on your mind.'

'Is this an interrogation now? How nice. I should have let you leave.'

'Korin told me how you and Adamov met in Leningrad, Maria. About your childhood. The siege.'

'So you know all there is to know about me then.'

'I'm starting to realize how little I know about you. For instance, I don't know why you had an affair with Fyodor Petrov.'

Her face became hard.

'I don't know what you're talking about.'

'On the roof, you called me Fedya.'

'Those pills turn you inside out. I must have been raving.'

Vasin saw steel in her gaze, cold and steady.

'So you and Petrov were never lovers?'

'Believe what you want. You already have your theories.'

Maria's gaze was defiant. A flash of the ferocity he had seen on the roof of the Kino the previous night played across her face.

'That's not "no".'

'Are you trying to frighten me, Major? If you think you can use what happened at the Kino against me, you can't. My husband will not be shocked by anything you can tell him. This is none of your business.'

'This is precisely my business. Fyodor Petrov dies under, let's say, unusual circumstances. The last people to see him alive were you, your husband the Director, and Colonel Korin. He was your lover, and a week later, you swallow a bunch of pills and try to jump off the roof of a cinema.'

'You're wrong. It's private and . . . unconnected.'

'Fine.' Meaning – enough.

Maria had broken off her gaze and was staring down at her own stockinged feet. When she looked up again, she had rearranged her face. She was suddenly calm, absolutely in control of herself, her anxiety wiped away without a trace. They might as well have been discussing the weather.

'Thank you, Major Vasin, for what you did last night. I still believe that you are a good man. It's good to have someone to talk to. I don't, you see. All the men in my life have their great deeds to accomplish. No time for the likes of me. Which is fine. But . . . Do you talk to your wife, Major? I mean really talk. You do have a wife, don't you?'

He began to answer, but she reached up and covered his mouth with her small hand. It smelled of medicinal spirit.

'Let's do something normal tomorrow. Not complicated. Not clever. Will you take me for an ice cream in Lenin Park? I can be there at eleven.'

It took a second for it to register. *Will you.* The familiar form. Something very simple, but also shockingly intimate.

'Yes.'

This time she allowed him to stand and leave. He let himself out.

It would not be normal, Vasin knew. Or uncomplicated.

II

The light in the living-room windows, bright and artificial as orange soda, proclaimed that Kuznetsov was at home. As Vasin mounted the stairs, he heard the record player cranked up, filling the stairwell with a plaintive Ukrainian folk melody. He opened the door into a pall of cigarette smoke. Kuznetsov sprang to his feet, spilling the contents of a full ashtray onto the carpet.

'For fuck's sake, Vasin! Where the hell have you been?'

'Good evening to you too. I hopped on a plane to Olenya.'

'You *are* joking.'

'I wanted to see Korin.'

Kuznetsov's hands flew to his temples in a pantomime gesture of incredulity.

'Without permission?'

'I thought Adamov's authority was all that one needed in Arzamas.'

Kuznetsov made a strangled sound of exasperation.

'*Don't.* Don't even move. Stay there.'

Kuznetsov picked up the telephone and dialed a four-digit number, his eyes fixed on Vasin as though he might disappear once more if he took his eyes off him.

'He just showed up . . . Olenya . . . Yes. *Olenya* . . . Leave my mother out of this, Efremov. Yes, I will. Of course . . . Understood.'

He replaced the receiver with an exhausted sigh and flopped back down on the sofa.

'You.'

'I know. I didn't mean to get you into trouble. What did Efremov say?'

'Nothing repeatable. Zaitsev wants to see you. Tomorrow.'

Vasin leaned on the doorpost to tug off his tall service boots.

'Got any dinner for me, old man?'

'Help yourself.' Kuznetsov sighed. 'The *kontora* won't let you starve.'

When Vasin emerged from the kitchen holding a deep bowl of steaming *solyanka*, he found Kuznetsov studiously absorbed in a pile of papers that covered the coffee table. He settled into one of the armchairs and ate hungrily.

'Good soup.'

'General Zaitsev will be so delighted to hear you're enjoying our hospitality.'

Vasin continued spooning soup and eyed Kuznetsov steadily until he deigned to look up.

'Whatever you think you are up to, Vasin, it's not going to work. Not the way you're doing it.'

'What exactly am I doing?'

'You are not in Moscow. You can't just disappear. This is Arzamas, six days before the biggest nuclear test in history. The *kontora* is going nuts over security. The Institute is a madhouse. Nobody is sleeping. The whole city's burning truckloads of fucking lightbulbs, working round the clock. And then there's you. Barreling around, pulling the top brass away from their jobs. Trying to start some half-assed murder investigation. Hitching rides to Olenya. I mean, what planet are you on, Vasin? How do you think this bull-in-a-china-shop act is going to help you? You'll just get yourself shut down, and pronto.'

'Zaitsev wants to shut me down anyway. Today was meant to be my last day, as you know.'

Kuznetsov did not reply, but instead scooped his papers aside, revealing a pair of heavy shot glasses. He fished a foreign-looking bottle out from under the table.

'Drink? Rum from Cuba. All the rage since Comrade Castro's visit.'

He poured two glasses, brimful.

'Our Latin comrades drink it with the juice of limes and coconuts. But until we get a fraternal delivery of those, we'll have to drink it our way.'

Kuznetsov raised his glass. Vasin, after a moment's hesitation, followed suit.

'To us.'

Kuznetsov breathed out sharply, made his mouth an O, and knocked back the liquor.

'Good stuff. Sorry, again. If I got you into any trouble.'

'Ah, Vasin. Don't say things like that. I'll start to worry you're doing something you shouldn't.'

'I should have told you about Olenya.'

'No, you shouldn't have. Not if you wanted to ever get up there. I would have had to tell the *kontora*, and they would have stopped you. But you already knew that.'

'I knew that.'

Kuznetsov poured another round. Vasin noticed that half the bottle was already finished.

'Do I feel a serious man-to-man conversation coming on so you can advise me to go home?'

Kuznetsov ignored the question. He pointed at the shiny Melodiya record player.

'Be a good sport and turn it over. It's the music from *Evenings on a Farm near Dikanka*. Great film.'

'I haven't seen it.'

'Excellent adaptation.'

'Are we going to talk about the short stories of Nikolai Vasil-yevich Gogol this evening?' Vasin returned to his chair and picked up his brimming glass. 'That would be most pleasant.'

Kuznetsov grinned, gestured silently with his glass, and knocked it back.

'So what did our lugubrious friend in Olenya have to say? I've always found Colonel Korin rather mysterious, myself.'

'He told me to go fuck my mother.'

'That may even be true.'

'Hasn't got much time for the likes of us Chekists, as he puts it.'

'Ah. Of course.'

'Of course what?'

'There's not much love lost. You can understand. So many of them . . . you know.'

'So many of them *sat*?' Vasin didn't need to finish the phrase. *Sat* always meant *sat in jail*.

'Right.'

'You knew about Korin?'

'He has that look, if you know what I mean.'

'Who else from the Citadel was in the Gulag?'

'I don't think Fyodor Petrov spent much time felling trees in the Arctic.'

'Be serious, Kuznetsov.'

'I don't know. Honestly. You'd have to visit the library and look it up.'

Kuznetsov winked theatrically.

'Good to know you're keeping abreast of things.'

'Did it happen to any of yours?'

'*It?*'

Kuznetsov and Vasin both sat forward, facing each other like chess players across the coffee table.

'The repressions. The purges. Was anyone from your family arrested?'

Kuznetsov shook his head.

'No, nobody? Or no, you don't want to tell me?'

'Both, I suppose. Nobody ever asked, apart from the personnel department when I joined up. How about yours?'

'My wife's grandparents. *Razkulachevenny.*' Meaning they had been arrested during the campaign against wealthy peasants. It had happened to so many that it had become a verb.

Kuznetsov grunted indifferently.

'What the devil are we talking about this for?'

'I don't know. Maybe because nobody ever talks about it. Maybe because it's still important to some people.'

'Nobody talks about it because it's ancient history. Five years since we played out that historic blame game. Khrushchev's big speech.' Kuznetsov slipped into the General Secretary's distinctive southern peasant drawl: '"Comrades, the Party made a mistake. There was some overzealousness in the elimination of enemies of the people."'

Vasin cracked a smile at the outrageous irreverence, but Kuznetsov's voice became urgent and confiding.

'He was right. It *was* understandable. We were fighting for survival. Deadly enemies were all around, determined to sabotage our glorious October Revolution. Some innocents suffered. Regrettable. Investigations were undertaken, thousands of victims rehabilitated. And pardoned, posthumously. Soviet justice has been restored. Our worthy leaders have cleansed the record. That is their gift to us. Our generation is blameless, and the older generation guilty. And the only men who are called upon to atone for their sins lie safely in their graves. Some of them in the same mass graves as their victims. Some of them in the Kremlin wall. Case closed. We are free of guilt, free of the past, free to build the future. Why dig that up?'

'I don't think Korin is free of the past.'

'I thought he didn't talk to you.'

'Just got that impression.'

Kuznetsov snorted and flopped back on the sofa. He fumbled for a cigarette and blew smoke at the ceiling.

'You're amazing, Vasin. I'm watching with bated breath to see what you do next.'

'Nice to have an appreciative audience.'

'First you make it obvious that you're planning to turn the Petrov case into a murder inquiry. Then you take an unauthorized jaunt to one of the most sensitive military installations in the Motherland. *On* the day of a test flight. Like you're some secret agent sent to spy on the program. Then you start asking questions about who sent who to the Gulag. Where do we take things from here? Start asking around about who's fucking the General's wife?'

It took Vasin's rum-addled wits a second to work out that Kuznetsov must mean Zaitsev's wife, not Orlov's. Zaitsev's *wife*. There was a terrible thought. Vasin reached out to pour the last of the rum.

'What did you mean, "who sent who to the Gulag"?'

'You know what I meant, Vasin. Everyone in that generation denounced each other. Kill or be killed. Wolves' laws. That's how it was.'

'And you don't think that kind of betrayal can echo down the years?'

'Maybe it can. But it's not our job to listen to echoes.'

'I thought we listened to everything.'

'I can tell you what *my* job is. To secure the future of our country against our enemies. Or don't you think that we have enemies?'

'We have enemies.'

'And traitors? There are no traitors? How about saboteurs?'

'Don't speak to me like I'm a child, Kuznetsov.'

'Okay, you're not a child. But you're an idealist. You're pursuing the truth, as you see it, without regard for the consequences. I call that naïve. And dangerous. I told you what the stakes were. The jeopardies that threaten the work here. Zaitsev and the military goons waiting to tear the cloud dwellers down. So whatever it is you think you are doing, for God's sake remember that. Don't do the bastards' work for them. The past is gone. Let it lie.'

Vasin and Kuznetsov sank back into the soft upholstery, exhausted as an old couple after an argument. Both lit cigarettes and smoked in silence. The only sound was the soft, rhythmic hiss of the record as it turned on its endless inside loop.

'What was the Kharkov Institute of Physics and Technology?'

'Vasin, go to hell. You're impossible.'

'We're on the same side.'

'If you say so.'

'You can help.'

'Help you put someone in jail for Petrov's murder and put another bar on your collar? Do you mind if I don't?'

'I understood what you said about the Citadel. I'm not a fanatic. But it's more complicated than you think.'

'Mate. Can you do me one favor? One? I don't want to know. Really. Please, keep your fucking *complications* to yourself.'

CHAPTER SIX

WEDNESDAY, 25 OCTOBER 1961

FIVE DAYS BEFORE THE TEST

I

Vasin rose through layers of sleep like a thrashing diver, struggling upward through water. He awoke tangled in unfamiliar blankets in a strange room to the rising wail of a siren.

'Kutuz . . . Kuznetsov!'

He skidded into the kitchen, his bare feet sliding on the smooth linoleum. He caught himself on the doorframe.

'Slow down, Young Communist. Don't break your neck. It's only an air raid.'

Kuznetsov's voice came from inside his room, from which he emerged a moment later with unlaced boots and a winter overcoat over flannel pajamas. In his hand he held a sheaf of notes and a dog-eared reference book.

'I mean, only an air-raid *drill*. Grab the milk from the fridge. And put on an overcoat.' Kuznetsov's voice receded as he trudged down the stairs. 'Bring your passport, and something to read.'

Vasin tugged on his tracksuit and coat and followed Kuznetsov down the stairs at a run. He caught up with him on the ground floor.

Kuznetsov stood holding open a heavy steel door leading to a cellar.

'The patrol should be around within the hour to check that we're all safe and sound. Come down and meet our lovely neighbors. They're already downstairs, taking cover from the Imperialist aggressors.'

The shelter was a low basement, equipped with rows of neatly made-up bunk beds around the walls. A dozen men and women, motley in dressing gowns and fur coats, slumped in easy chairs and sat at a pair of tables, slumbering or reading. A few glanced at Vasin without particular curiosity. In one corner a pair of young twins laid out a desultory game of checkers. Kuznetsov, acknowledging nobody, bounced down himself into an easy chair and cracked open his book. Vasin sat alone on a bunk, regretting having given away his copy of *Krokodil*. The group in the cellar, settled in their various attitudes of study, seemed to him a bizarre parody of a university library reading room on a quiet afternoon.

The banging on the door was curt and officious. Eyes turned to Kuznetsov. With a sigh he rose and unrolled the door bolt.

'Greetings, early-rising Comrades, we have the roll call ready here for you . . .'

Kuznetsov paused, midsentence. The door opened to a pair of young sergeants with KGB epaulets. The elder one saluted.

'Good morning, sir. Major Kuznetsov?'

'Okay, okay. I know who sent you. Back upstairs, Vasin. We're getting dressed. There's somewhere we need to be.'

The uniform suited some men, fitting as though they had grown up in it. Kuznetsov was not such a man, Vasin decided as he clattered down the stairs behind him. In his breeches, tunic, and pistol belt, Kuznetsov looked like an actor in a provincial theater stuffed into a badly fitting costume. Their breath steaming in the morning chill, they scrambled into the back of a UAZ jeep for the short drive into town. The sirens' wail swelled and ebbed, relentless, as they passed the speakers mounted on each lamppost. Lenin Square was deserted except for a pair of street sweepers' handcarts, abandoned by their owners. A tram stood by the intersection with Kurchatov Street, empty, except the driver had forgotten to extinguish its single, cyclopean headlight.

Instead of turning in to the *kontora*'s forecourt, they slewed past, drove on to another square official building that stood on a high bluff overlooking the old monastery.

'*Kommandatura*,' Kuznetsov mumbled by way of explanation.

'Military command headquarters. The lair of General Pavlov. Head soldier around these parts. Runs everything in this city – outside the walls of the Citadel, that is.'

That familiar, masculine smell of military installations: floor polish, new uniform cloth, strong tobacco. Vasin thought of the Army training camps he and his fellow students had been made to attend while he was at university. Cold showers, mud, bawling voices. The Army's endless hierarchy of petty cruelties, the king-dom of the stupid. They followed a group of infantry officers hurrying up a broad concrete staircase to the top floor.

Around forty men stood in groups in a large, carpeted anteroom, speaking in low voices. Most wore Army green, with a scattering of KGB blue. As Vasin and Kuznetsov joined them, a set of double doors at the opposite end of the room swung open. A heavyset man lumbered in, carrying a teacup. Everyone snapped to attention. General Pavlov ignored their presence. He sat down heavily at a long conference table and took a slow swig of his tea.

Pavlov glared at his wristwatch and then at the clock on the wall as though daring them to contradict each other. Half past six.

'Enough.'

His voice carried effortlessly. A young orderly acknowledged the order meekly and hurried out. Within a minute the wailing sirens ceased, leaving a silence that rang in the ears. Pavlov decided that he needed to drain the last of his tea, and needed the room to watch him do it too.

'At ease.' The words came from deep in Pavlov's chest; he could have been clearing his throat. '*Nu?* Well?'

An adjutant stepped forward nervously, neat in a freshly pressed tunic.

'The air-raid test was a success, Comrade General. Antiaircraft crews at their posts in good order. All but one of the new emergency generators started and are running. We have had phone reports from all the shelters. Ninety-four percent of the civilians accounted for and at their muster stations . . .'

'The civilians. Where's our Comrade Major General Zaitsev?'

On cue, the sound of boots on the stairs. Major Efremov stepped in smartly, followed by Zaitsev, puffing from the long climb. The crowd parted before him. Vasin sheltered from the General's roving, furious eye behind a tall Air Force captain.

'General, good of you to join us.'

Zaitsev settled wordlessly into a chair beside Pavlov.

Efremov stepped forward, opening a leather folder and proffering it to the commandant. Pavlov fixed the younger man with an out-from-under stare of disdain, taking in his polished boots, pomaded hair.

'Six percent of our little pigeons unaccounted for, Zaitsev.' Pavlov raised his eyebrows along with the tone of his voice. This, Vasin imagined, indicated sarcasm. 'Decided not to get out of their soft beds?'

'The KGB will make inquiries, Comrade,' Efremov began. 'But as you are aware, the members of the Institute are not always able to immediately . . .'

The adjutant's voice trailed off as Pavlov hauled himself up and walked over to a set of tall windows that filled one end of the room. He looked like nothing so much as a side of beef in a uniform: the skin of his face and hands blotchy red, his neck straining against the collar of his tunic. Gray hair stuck up at the top and was shaved short at the back and sides. On his uniform four rows of medal ribbons were punctuated by the single, unmistakable gold star of a Hero of the Soviet Union, on its scarlet ribbon.

'And what is this?' Pavlov pointed curtly.

Obediently, the officers turned to look across the landscape of rooftops that spread below them. One large glass roof was illuminated from below, the glow of the fluorescent tubes fading into the spreading dawn. Even Vasin could see that it belonged, undoubtedly, to one of the workshops of the All-Union Scientific Research Institute of Experimental Physics.

'Major?'

Pavlov ground a fat finger into the glass of the window in the direction of the illuminated laboratory, then tapped it for emphasis, leaving a greasy mark.

'A shocking breach of regulations, sir.' Efremov's voice was theatrically shrill with indignation. 'It seems to be the new laboratory complex. I assure you the duty security officers were fully briefed on the revised blackout procedures.'

'And the boys in white coats told them to get lost. To them, this is all provided for their amusement! This whole establishment, so that they can live in the clouds! Is this negligence? Or sabotage?'

'We will find those responsible immediately.'

Pavlov tapped the window once more, for emphasis, then stalked out of the room. Zaitsev remained silent, arms crossed over his chest, his face a brewing storm. For a long moment nobody moved.

'Well? What are you all standing around for, like heifers waiting for a bull to fuck them?' Zaitsev bellowed. 'Dismissed!'

Vasin and Kuznetsov joined the rush for the door. By the time they had reached the ground floor, most of the Army men had peeled off into their offices, leaving only their KGB colleagues crowding into the cloakroom like disgraced schoolboys.

The day was fully born now, a pall of low cloud promising more snow. After the overheated fug of the *Kommandatura*, the morning air felt cold and clean on Vasin's face.

'So that was the charming General Pavlov.'

Kuznetsov rolled his eyes as he buttoned his overcoat.

'Zaitsev's even more evil twin? Hero of the Soviet Union, my ass,' Kuznetsov answered in a low voice. 'That Pavlov. Desk jockey. Politburo ass-kisser. Heroically cleared out all the old warhorses who'd actually bled on the front. Pavlov's war was about filling tank shells and fucking women munitions workers with tits like torpedoes. That's what they say down at the *kontora*. But sometimes the meat-head in green likes to show everyone that it's the military that ultimately runs the show around here.'

'Vasin!'

Zaitsev had emerged at the top of the steps, punching his way into his overcoat as though the flapping cloth deserved a good thrashing. Vasin trotted back up and saluted smartly. Efremov, hovering at his boss's side, wordlessly handed Zaitsev a piece of paper. Vasin recognized the list of movement orders from the previous day. The paper crumpled in the General's fist as he thrust it forward.

'I never authorized a flight to Olenya.'

Vasin caught a whiff of the General's hungover breath. As he looked up into Zaitsev's porcine face, he noted the unshaven chin, a corrugated pattern of corduroy indented on the pale flesh. A night on the sofa? He bowed respectfully before the General's halitosis.

'I was interviewing an important witness, Colonel Korin. Wanted to get the investigation done as quickly as possible. As per your orders, sir.'

Zaitsev snorted like a bad-tempered bull and took a step down, his face now level with Vasin's. His voice was low and menacing.

'You were meant to be gone yesterday, but you're still fucking here. You're a hemorrhoid I can't get rid of.'

'I . . .'

'Your *orders* do not trump what's going on here. You're in Arzamas, not in some Moscow whorehouse. And however big you think you are, Major Vasin of State Security from Special fucking Cases, however big your boss Orlov thinks he is, no one is bigger than what's being built here in Arzamas. So go, run and bleat to your chief. Get your permissions from on high. But I swear to you. I *swear* to you, Vasin: Step out of line one more time, disappear one more time, and I will break you. Officially. Unofficially. Whatever it takes, I will do. But you will play by my rules. My. Fucking. Rules.'

Vasin paused to slowly wipe the flecks of Zaitsev's spittle from his face. He stood to attention and saluted. As he walked down the steps away from the glowering General, Vasin felt his bowels turning to water.

II

Orlov's private telephone network was one of Special Cases' more ingenious secrets. Though of course it was not technically private, only a private use of something very public. 'We hide in plain sight, Major,' Orlov had told Vasin when he briefed him on the system. 'The most private place for a conversation can be in a crowd. The most secret place for a message may be on the front page of *Pravda*.'

The USSR's railway network had its own telephone and telegraph system, independent of the All-Union Post and Telegraph, which connected all other calls. The secret was that Orlov had his own patched-in line, accessible from any railway station across the country. The network was antiquated – some of it even pre-Revolutionary – but secure for precisely that reason. No one in the *kontora* had much interest in listening in to the conversations of stationmasters and junction managers. A parallel phone system that covered the entire Soviet Union.

Vasin parted from Kuznetsov on the corner of Lenin Square and boarded a tram toward home. This time his companion made no protest, which surely meant that Zaitsev had made other arrangements to keep tabs on him. To give Vasin space to make a mistake. Sure enough, Vasin spotted two men in plain coats scramble aboard as the doors closed, studiously avoiding his eye as the tram gathered speed. Giving them the slip should be easy. Shaking them off so they didn't realize they'd been shaken would be harder. Back at the apartment Vasin changed into his civilian clothes and stuffed a black beret into the pocket. He returned to the *kontora* on foot, watching his followers out of the corner of his eye. Once inside the building he nonchalantly made his way to the cafeteria and, as he waited in the queue to pay for his sandwiches, scanned the room. His shadows had remained outside.

A back door, a courtyard, a low wall behind a line of rubbish bins. The service gate stood open as a delivery truck backed crookedly out into the street. Vasin slipped past and worked his way back around the building. Through the dripping trees he spotted his watchers waiting docilely for him in a Volga sedan.

Two more trams, boarded and hopped off at the last moment, brought him to Arzamas's small train station. Vasin found the operations room without difficulty. The duty stationmaster obeyed Vasin's scarlet KGB identity card and showed him to a chunky iron telephone receiver that hung in a corner.

'We don't use it much these days.' The stationmaster was a small, anxious hippo of a man, eager to please. 'This branch is all electronic now.'

He pointed to a large wall-mounted board showing the Mid-Volga Railway Region dotted with red and green lights. A pretty, silent blond secretary hovered protectively close to her boss. Vasin's professional eye caught the body language of an affair. Possibly recreational humiliation? He smiled at his own jadedness before nodding both back into their offices. What have you become, Vasin?

He dialed the four-digit number that got him a priority line to the local railhead: the city of Gorky. Another code, and he was connected to the trunk line to Moscow's Yaroslavsky Station. Four tinny rings and an operator picked up. Raising his voice to be heard above the static, Vasin asked for Moscow Heavy Goods Ring Line Station, Number 262. Orlov's Lubyanka office.

After a short wait Orlov came on the line, the loudest voice in a whispering cacophony of railway men's chatter.

'Comrade! Good to hear from our fellow workers in the middle Volga! No problems? Traffic flow in your section good?'

'Nothing unexpected, sir.' Vasin had to almost shout; it was like trying to make himself heard in a crowded cocktail party. 'The usual problems. The head of the local organization. Unhelpful. Very unhelpful.'

'Manageable?'

'Yes, sir. So far, manageable.'

A deafening electronic buzz made Vasin hold the receiver away from his ear for several seconds.

'Still with me?'

'Still with you, sir. Requesting a routine information check. Code 111.'

The highest urgency.

'I am listening.'

'Two files. Both 1937. First case, subjects: Comrades Matveyev. Markov. Adamov.'

Orlov's simple security code to throw off any eavesdroppers, the first two names on any list were always to be nonsense. Then the real name. God loves trinities, Orlov had said.

'Second case: Comrades Ivanov, Sidorov, Petrov.'

'The last again please, Comrade?'

'Petrov. *A. V.* Petrov.'

'I hear you clearly and understand. And what information do you require?'

'A connection between the two cases. The comrades in Case One suffered a . . . serious derailment. In 'thirty-seven. I believe that the individuals in Case Two may have been responsible.'

'Responsible for the derailment? We will check our records.'

'My thanks.'

'Your diligence does you credit, Comrade. Call tomorrow. Same time.'

Vasin breathed a sigh of relief as he replaced the heavy steel receiver. Orlov's voice had been friendly. Vasin's wife was evidently still holding her tongue. No sign that the General had any idea about him and Katya.

III

Vasin glanced at his watch. He judged that he probably had a couple more hours before his minders began to grow suspicious.

A standard-issue statue of Lenin stood at the entrance to the main city park, his concrete face resolute as he gestured toward the glorious future. Vasin hurried through the concrete archway, past a forlorn row of refreshment stands shuttered for the winter. Their brightly painted fronts were decorated with cartoon animals. The signs said, SODA, DOUGHNUTS, COTTON CANDY. Nikita's favorite. The Soviet city park, the space allotted to approved leisure activities. Walking. Eating sweet foodstuffs. Enjoying nature, tamed and framed and shorn of its wildness and hostility and laid out in carefully measured blocks, like a green model city. Also the designated space for all human activity not connected with labor. Courtship, for instance. Lovemaking in the spiky, pungent undergrowth. Strolling with infants. Talking to one's children. A place where Soviet citizens were allowed to be solitary, though never truly alone.

It occurred to Vasin that his walks in Gorky Park were probably the closest he had ever come to spending time in private with Nikita. Had he really ever had a real conversation with his son? Whenever they were together, without the nagging presence of Vera, it had always seemed kinder to allow the boy the luxury of silence. Every time he asked his son a question, even a joking one, the boy's face would pinch in earnest nervousness as he considered what would be the correct answer. The poor kid seemed always convinced that he was somehow at fault, but without ever knowing why.

An alley of birches gave on to a wide lawn, gray with melting snow and crisscrossed with footprints. By an empty bandstand Vasin caught a flash of electric blue, the only vibrant color in the monochrome landscape. Maria. From a distance, he watched her. On the street, in a crowd, she had walked upright and stiff as a windup doll. Now, when she thought she was alone, she kicked her boots through the snow like a child. Her hands were stuffed deep in her pockets, and her sharp, small face was almost hidden by the hood of her mackintosh. There was something about her fragility that sent a sudden, unexpected stab of protectiveness through him. He thought of her struggling, the strong kick to his face, as she had fought toward her fall the other night. The anger in that small body, the force of it. He walked out across the snowfield.

'Hi.'

'Hi yourself.'

He waited for her to continue, but she was back to the formal form of address. Vasin felt obscurely hurt.

'Aren't you going to buy the girl an ice cream?'

The cafeteria overlooked a round, concrete-lined pond, black under the white sky. A few grandmothers sat in the coffee-scented warmth, minding babies in prams who were wrapped like parcels. He bought a couple of ice creams, the forty-eight-kopeck variety he had always loved, a square block of vanilla-flavored cream. They walked on, eating in silence. There was something about eating ice cream in the snow that Vasin had always found pleasingly strange. It was like swallowing winter.

'Good. Thanks.'

Maria paused by a rubbish urn to lick the last of the ice cream from its wrapper, then deposited her rubbish like a good citizen.

'I come here quite a lot. Never with company, though.'

'You're lonely?'

'Always straight to the point. Maybe I always have been. What of it?'

She stopped and turned toward the maze of birches, her face hidden under her hood. Vasin let the silence between them grow.

'They're all crazy. You know that, right?'

'Who?'

'The people here in this city. The power they command and the secrets they keep twist them. It's hard to explain so that you might understand.'

Not so different from the rest of the world of the *nomenklatura* – the Soviet elite – Vasin thought. His year in Special Cases had been an education in the distorting power of privilege. How it can corrode men. And women.

'Was Petrov twisted? Adamov?'

'My husband is a good man.'

'If you say so.'

'Korin told you about Leningrad.'

'He told me some.'

'Where were you during the war?'

'Korin asked me exactly the same question. Does it matter?'
'Oh, yes. It matters.'
'Moscow. I was still at school.'
'So was I, until they closed the schools. No heat, no food, teachers all mobilized; some of my classmates made it out on the boats across the lake. To Ladoga. My father wanted to get me on the transport, but my mother said it was too dangerous. She was right. The Germans dive-bombed one of the ships. The water was full of little kids' summer hats, floating back to the shore. My friend's mother saw it. She drowned herself the next day. You didn't see the war.'
'No. Not the way you did.'
'That siren, this morning. Second time in as many weeks, for fuck's sake. Last time was on the morning Fedya died. I hate it.'
'Aren't you used to it by now?'
'No, because one day it won't be a drill. I think of that day a lot. The sky filled with bombs tumbling from the bellies of airplanes. Detonations like great doors slamming underground. The hard, solid things of the world, melting. Bricks, weightless, flying upward. Landslides of masonry and plaster. Buildings bursting like paper bags. Fire. Hot wind. Every time I hear that siren I think, This is the day we will all be erased from the world along with our death-breeding city. Every time I hear the siren, you know what I do? I put a pillow over my head and I wait. And then no planes come. Only another day to get through. Thinking about how that fucking siren was probably the last thing that Fedya ever heard.'
'You don't go down to the shelter?'
'I'll never go into a shelter again in my life. My neighbors hate me for spoiling their quota. Svetlana Ivanovna and her foghorn voice booming in the stairwell, exhorting all comrades to hurry to the basement. She never dares to knock on our door, but I can feel her dirty look as she passes. Her and the old cows muttering at the bread shop. "Arrogant little bitch," they say. "Whatever does the Director see in her?" Hypocritical, cock-sucking cunts, every one of them.'

Vasin started involuntarily at the obscenity. Maria's grimace of hatred turned into the beginning of a smile.

124

'I kept some bad company as a kid. It still comes out sometimes.'

'Sorry. Korin said you met Adamov in Leningrad.'

'He taught me. Pure mathematics. We both loved numbers. They're hard, beautiful, concrete things that can't ever be destroyed. All those infinite patterns, fixed until the end of time. Whether there are people to know it or not. We took comfort.'

'In each other?'

'In science. After what we had both been through. But yes. You're right. We took comfort in one another. I was seventeen when we met. An orphan, by then. Street-smart, but still more kid than woman. Scarecrow, Adamov called me. "A pair of big green eyes up top, a pair of tough little fists in the middle, oversize man's boots down below." I nearly punched the old goat when he tried to lay a hand on me.'

Maria smiled at the memory. Vasin suppressed a smile of his own at the thought of Adamov in the improbable role of old goat.

'He said, "I will look after you." And he has been as good as his word. My savior, I thought at the time. Like I told you. He is a good man.'

'And what happened on the roof of the Kino?'

'You're bloody relentless, you know that?'

'It was pretty memorable. For me, anyway.'

'What part?'

There was mischief in her glance. Vasin thought of her hands gripping his face, the smell of her skin, and looked away.

'You tried to jump. Why?'

'Maybe I wanted to fly away. Escape.'

'Is that why you had an affair with Petrov? To escape?'

Masha turned to Vasin, defiant.

'You tell me something, then I'll answer you. Do you think someone killed Fedya?'

'Maybe.'

'You've spoken to his colleagues?'

'Of course.'

'To Vladimir Axelrod?'

'Maybe.'

Masha's face creased in distaste.

'Axelrod's a pederast.' *Pederast*, criminal jargon. Masha spat the word. 'I mean literally. He is a homosexual.'

'How do you know?'

She stopped walking, her eyes fixed on the snowy ground.

'Because Fedya was sleeping with him before he started sleeping with me. I told you, this place is full of deviants. If you think someone poisoned Fedya, start by asking his jealous lover boy.'

Masha turned abruptly and walked away.

IV

Vasin picked up his watchers once more, striding nonchalantly out the front door of the *kontora* without a sideways glance. Like dutiful dogs, the pair slipped into position behind him as he walked to the Institute. A mist was thickening, and the sky promised a new snowfall.

Vasin watched Adamov and his acolytes file out of the lecture theater. The Director's eye slid over Vasin's shabby civilian raincoat without noticing him. Axelrod trailed the group, walking alone. He and Vasin spotted each other at the same moment. To Vasin's surprise, Axelrod pushed his way back through the crowd of young scientists. He was agitated.

'Major, I need to talk to you in private. I didn't know where to find you so I called the State Security headquarters and . . .'

'They said they'd never heard of me.'

'Exactly.'

'I'm only a visitor here. But, Doctor, I was about to say just the same to you. We need to talk.'

Axelrod's office was lined with orderly files and dominated by an outsize blackboard, which was covered in calculations. Nervously he swept a pile of papers aside and plumped a large ring binder marked TOP SECRET on the desk. On the cover the date – 25 OCTO-BER 1961 – was stamped in thick black ink.

'Major, this is the latest design of RDS-220. We get a new version every week, updated with all the new parameters. Every department's work, summarized, so everyone knows what everyone else is doing – the engineers, the metallurgists, the meteorologists.'

Axelrod began to leaf through the pages covered in figures, graphs, and engineering blueprints. 'Here, the conventional explosives experts and the fission boys. And of course Adamov's team, all the new yield projections of the main thermonuclear device . . .'

'Axelrod, are you crazy? This is classified!'

'So what? You could never understand a word of it.'

Axelrod registered Vasin's look and paused.

'Oh. Sorry. Was that impolite? I do that sometimes. I'm rude to people, without meaning to be.'

'No, you're right. There's nothing here I would understand. Go on.'

Flustered, Axelrod resumed his search and smoothed the file flat on the table.

'Right here, the section on the casing is, was, Petrov's work. It's all gone. He rewrote it.'

'Who?'

'Adamov did, just days before the test. He's reconfigured the whole damn apparatus.'

'Aren't engineers always changing things? Why do I need to know this?'

Axelrod snatched at his rumpled hair in a gesture of confusion.

'Because the casing is everything Fyodor worked for. We debated its composition for months. Fyodor was passionate, adamant. And now that he's gone, his work has gone too.'

'And you find this sinister?'

'No, I find it suspicious. You do not?'

'You are accusing Professor Adamov of being somehow complicit in Petrov's death because of a disagreement over' – Vasin gently closed the folder and pushed it back across the table – 'metallurgy?'

Axelrod deflated.

'I accuse nobody.'

Vasin paused. What Axelrod was saying might be important. But first he had to test if Masha was telling the truth about his relationship with Petrov. Gently, though. He didn't want to spook Axelrod into silence. But if what Masha said *was* true, it was the moment to put Axelrod on the hook.

'Before we discuss this further there is something I wish to raise with you. As I said, I actually came here to speak to *you*, Dr Axelrod. I have some questions about your relationship with Petrov. Your close friendship.'

Some men were hard to read, their countenances stone. Axelrod's was an open book. Alarm passed across his grief-stricken face. Ask any interrogator – sitting is an eloquent business. Suspects sit according to the guilt they carry, though not always, as Vasin had learned, the guilt that you are looking for. They would sprawl and straddle, fidget, cross and uncross their legs. What they never, almost never, did was sit in a posture that was finite and irreducible, not a muscle stirring. Yet here was Axelrod, frozen to the spot as though posing for a photograph. His long-fingered hands lay immobile on his thighs, his whole body suddenly halted in its small motions. But his was not the stillness of calm, it was the paralysis of fear. When he spoke, Axelrod's lips barely moved.

'My close friendship?'

'Is that not an accurate description?'

Axelrod went pale. Vasin pressed on, choosing his words delicately.

'Intimate friends.'

'Certainly not.'

'Relax, Comrade. I don't care about Article 121. We leave locking up people like that to the cops. If you ask me, it's a pointless task.'

Axelrod found motion, suddenly, like a paused film set once more to run. He sat forward, and his hands leapt at each other like a pair of fighting animals.

'It's an outrageous insinuation. You have no proof.'

'Forgive me, Doctor, but that is not an answer.'

'The suggestion is disgusting. Offensive. Who told you such a thing?'

'Does it matter?'

'There are many here with grudges. Men become obsessed.'

'Are you saying someone is pursuing a vendetta against you?'

'Fyodor and I are young . . . were young. We had authority above many who are more senior, but less able. I am speaking of jealousy, Major. Evil tongues. Malicious gossip. Such malicious filth in people's minds. My God, to accuse us of such revolting deviancy.'

It was not the first time Vasin had seen a man drawing strength from the passion of his denial. Axelrod, a moment ago startled as a rabbit, had found the mettle to raise his voice. He stood, abruptly, tipping his chair onto the floor.

'If you came here to threaten me, Major, with some ridiculous inventions for your own purposes, I can only say that I refuse to play your foolish games. I protest in the most adamant terms!'

Axelrod's passion petered out, battered flat against the rock of Vasin's silence. Slowly the investigator half-stood, finding a higher perch on the corner of the desk, and crossed his arms over his chest. Axelrod's eyes traveled across his face, looking to see if his performance had been believed. Involuntarily, the scientist's hands sought each other once more and clasped tight for comfort. He glanced down at the upturned chair, decided against stooping to pick it up, then faced his accuser once more. God, thought Vasin. So it's true. He felt an involuntary surge of pity for this evidently brilliant, brittle man, whose weakness had placed him suddenly in Vasin's power. And yet, Axelrod was clearly a man who held secrets. Perhaps the secret of Petrov's death.

Vasin made his face benevolent.

'Do you have a girlfriend, Doctor? If you will forgive the personal question.'

'My fiancée is in Moscow.'

'Well then. Malicious rumors. Anyway – like I said, I don't care. And you are right – I don't have any proof about your personal inclinations or about your relationship with Petrov. Yet.'

Vasin let the last syllable hang in the air for a moment. He continued, leaning forward, his voice low.

'All I care about is finding out who killed Petrov.'

'As do I.'

'We see eye-to-eye, Comrade Doctor. And I need someone inside the Citadel. Please. Sit down.'

Axelrod fumbled to right his chair.

'You want me to become your informer?'

'My guide. We have the same goal in this, we agree.'

Axelrod nodded bleakly.

'You want to put me on the hook?'

Vasin had always found the colloquialism for being recruited by the KGB imprecise. The *kontora* liked to land its fish immediately, with a single violent jerk. Then they would leave them to gasp, drowning in the air. Perhaps, then, they might agree to gently release the fish back into the water to swim a little longer, as far as the line would allow. The hook was already set deep in Axelrod's throat, even if he didn't yet realize it.

Vasin weighed his sympathy for this floundering, flawed man against his hard investigator's instinct. The policeman won. He decided to let Axelrod flap around a little more.

'You are not on the hook if you have nothing to hide, Comrade.' Vasin repeated the old secret policeman's lie too easily for his own liking. 'I am only asking you to help find the answers you seek yourself. You came to me voluntarily, remember?' Vasin placed a palm on the classified folder in front of them. 'You wanted me to understand something about metallurgy. Start there. Guide me in terms an idiot could understand. Why should I care that Adamov changed Fyodor's casing? Why did you want to tell me about it?'

A nervous smile flicked across Axelrod's face. Vasin had seen it before. The face of a man who has been momentarily dangled over the precipice and then hauled back by firm hands into the warm embrace of collaboration. The brief storm of alarm, the terror of discovery, had passed. Now Vasin gently released Axelrod back into his natural element, his deep sea of numbers.

'Where do you want me to start?'

'Start with how the bomb works?'

'It's basically very simple. You are familiar with the concept of an atom?'

Axelrod evidently found it hard to judge where the limits of a layman's ignorance could lie. Vasin nodded gravely.

'At the center of every atom is a nucleus made of two kinds of particles, protons and neutrons. A different number of protons makes a different element. Hydrogen, one proton. Helium, two. And so on. And around every nucleus are concentric rings of electrons, like little moons circling around a planet. When the electrons move from atom to atom, that's called electricity.'

Vasin half-expected Axelrod to ask whether he had heard of electricity.

'Most atoms are stable, meaning the nucleus has an equal number of protons and neutrons. But some elements, especially the heavy ones, have an unbalanced number of neutrons. Nature abhors disequilibrium, so they spit out their spare neutrons, naturally. When neutrons move, that's called radiation. If you leave a radioactive element alone, it will spit out all its spare neutrons and eventually become inert. We call that radioactive decay, so over time, uranium 235 will eventually turn to lead.'

'Uranium what?'

'Two thirty-five. It's a kind of uranium. A heavy metal. Most uranium is pretty stable and barely radioactive. But about half of one percent of naturally occurring uranium has a different number of neutrons from the normal sort. That's called uranium 235. It's more radioactive. And very unstable, because when you add just one more neutron, it becomes uranium 236. And that atom is too heavy to exist, so its nucleus immediately splits in half, into barium and krypton. When the nucleus splits, that's called nuclear fission. It releases enormous amounts of energy. Uranium fission, for instance, produces about eighty-three terajoules per kilogram.'

'That is a lot?'

'Compared to oxidizing hydrocarbons, I mean, compared to burning coal? About twenty-five million times more energy, roughly.'

'You measure the power of atomic bombs in coal?'

'No, actually. We measure the power of bombs in tons of high explosive. TNT, to be precise. It's not really a scientific metric, but it's a more useful indicator than energy content when you are measuring the impact of the device on the real . . . um, world. For instance the first American device, Little Boy, dropped on Hiroshima in 'forty-five, yielded eighteen kilotons. The same as dropping eighteen thousand tons of high explosives at the same time.'

'My God.'

'Little Boy was tiny by modern standards.' Axelrod straightened with pride. 'The detonator alone for RDS-220 is a fission bomb larger than Little Boy.'

'A detonator as powerful as eighteen thousand tons of TNT?'

'We talked about fission, splitting atoms. When you put a certain amount of U-235 in one place, the radiation, the neutrons coming off all those disintegrating atoms start to split each other apart. Like rolling a billiard ball into a cluster of other billiard balls. Free neutrons knock out other neutrons, which knock out others. That reaction also gives off heat, and more neutrons, which in turn split more of the surrounding atoms in a spontaneous chain reaction. That's what happens in a nuclear reactor. You put enough fuel in, and it starts to generate heat on its own.'

'You mean it starts to destroy itself spontaneously?'

'Right. But you can control the reaction by absorbing those free neutrons with graphite rods. It's like putting a dishcloth on the billiard table. Some of the balls will roll into it and stop knocking each other around. With graphite, we can control the reaction and stop a nuclear reactor from melting down. We've even been putting reactors in submarines. *Kit*-class attack subs to start off with, in 'fifty-nine. And last year we launched the first nuclear-powered missile sub. There have been some problems, though. Back in July there was a terrible accident on K-19 out in the North Atlantic – the reactor coolant leaked, and the crew have been dying like flies ever since . . .'

'For God's sake, Axelrod. Do you even know the meaning of the word *secret*? One more word about submarines and I'll have to bury you myself.'

Axelrod's face tightened anxiously for a second as he weighed whether Vasin was joking, then relaxed into a nervous smile. 'An atom bomb uses the same principle as a nuclear reactor. A fission reaction. But instead, it's designed to melt down on command, explosively. To do that you need to focus the reaction.'

'How?'

'If you just put a pile of blocks of uranium 235 together, they'll start to react as soon as you have a critical mass. Takes about fifteen kilos of pure 235. It will get hotter and hotter and emit more and more radiation. But it won't explode. You'd just get a kind of nuclear bonfire that would burn through the floor. And probably through the earth's crust. Nobody's tried it . . .'

Axelrod paused for a moment, as though considering how interesting such an experiment would be.

'And how does it go bang?'

'So, to get it to explode, we need to achieve critical mass very suddenly, and under great pressure. That means we have to enclose the reaction in some kind of vessel. That's called the tamper. And we need a way to transport the critical mass of uranium in a safe state until you're ready to detonate it. Actually, the answer is very simple.'

The scientist glanced questioningly at Vasin, as though at a particularly unpromising student. 'Obvious, really.'

Vasin spread his hands with a gesture that said, No idea.

'You cast the uranium 235 into a hollow ball.' Axelrod cupped his hands to imitate a sphere. 'The critical mass is all there, but there's not enough of it in one place to trigger a chain reaction. It doesn't go critical on its own because it's hollow in the middle, you see? It's too spaced out.'

'And then?'

'Then you surround this hollow ball with an outer shell of explosives. You detonate those explosives and the blast makes that ball implode.' He squeezed his cupped hands into a tight fist. 'Suddenly, it's a solid ball and therefore goes critical. A chain reaction kicks off and energy is released. Nuclear explosion.'

'That's what happens inside that?' Vasin pointed to the file on the desk.

'Yes. No. I mean, both. RDS-220 isn't like an old-fashioned fission bomb. It's a thermonuclear device. A hydrogen bomb, in common parlance.'

'Which is?'

'Very different. The first atom bombs worked by splitting heavy atoms apart, nuclear fission. Thermonuclear bombs work the opposite way. They make lighter atoms join *together*. Nuclear fusion. That's what happens in the heart of the sun, in all stars. They are all giant, continuous thermonuclear explosions. Balls of hydrogen, fusing together to make other elements and giving off light and heat. Every atom in the universe was created inside a star. Every atom in your body, in mine. To reproduce the effect on earth we need to create the same conditions as on the sun. We expose hydrogen to heat and pressure. Something like seventy-three million times the pressure of the atmosphere of earth. And the fusion begins.'

'And how do you make a sun . . . on earth?'

Axelrod smiled in fond pride, as though Vasin had asked him about the school grades of a particularly brilliant child.

'By using a fission bomb as a detonator, we get the necessary energy. As long as you surround the detonator and the hydrogen in a strong enough casing. The tamper has to be very thick and heavy to withstand the pressure for as long as twenty, maybe thirty milliseconds. Then you get enough energy concentrated inside to start a fusion reaction.'

'Which is explosive?'

'What you call an "explosion" is just rapid combustion. Combustion is something solid turning to gas and expanding. Expose almost anything to enough heat and it will turn into gas. So, when the hydrogen atoms in the bomb fuse together into different elements, they release enormous amounts of heat and light that vaporize everything around them. But that explosion also vaporizes the casing of the bomb itself, releasing the pressure, and the fusion reaction stops. That was the genius of Petrov's design. The heavier and stronger the tamper, the longer it can contain the fusion.'

'That can happen in twenty thousandths of a second?'

'Certainly. That's long enough to create a sun. A small sun. And you see, the stronger the tamper you have, the longer the reaction time. Petrov's idea was to make a casing of twenty tons of pure uranium metal. Very dense. Very strong.'

'Twenty tons of uranium? But you said that fifteen kilos was enough . . .'

'Fifteen kilos of weapons-grade uranium is enough to make an atomic bomb, yes. That's uranium 235. But like I said, natural uranium metal contains less than one percent of that stuff. Ninety-nine percent is uranium 238. It is much more stable, but has never been used to make a tamper. We always used lead. But Petrov was a genius, a revolutionary. He wanted to make RDS-220 out of uranium for two reasons. It's much denser than lead – about twenty tons per cubic meter for uranium, eleven for lead. And uranium's boiling point is more than twice as high.'

'Stronger tamper, bigger explosion.'

'Correct. Well done.'

'And the other reason?'

'Ah. Here is the pure beauty of Fyodor's vision.' Axelrod sat forward, his whole body animated. 'A new generation of devices, an order of magnitude bigger than anything ever seen before. He wants to open a new chapter.' He caught himself, as though stung. 'Wanted.'

'A new chapter of . . . ?'

'Petrov's idea was a three-stage bomb. The first two stages were a standard hydrogen bomb. A fission device as a detonator, then two chambers of hydrogen as the main explosive force. But the casing itself would act as a third stage. The neutrons thrown off by the fusion would irradiate the uranium tamper. Even the trace amounts of U-235 could become fissile with that much radiation and heat. So you have three stages: fission, then fusion, then once again fission. The uranium tamper was revolutionary. It would have doubled the power of the device. More, perhaps.'

'Might? Using a uranium casing has never been tried?'

Axelrod shook his head.

'So what did Adamov change?'

'The day after Petrov's death, Adamov ordered the metallurgists who were casting the uranium tamper to stop. And ever since he's been closeted night and day with his closest comrades recalibrating. And today – this.'

Axelrod traced a finger over the cover of the thick document that lay between them.

'The new tamper is made of lead. Not uranium, but lead. Which is inert. There is no third-stage fission. Professor Adamov's substitution of lead for uranium will drastically reduce the power of the device. It is the opposite of what we have been trying to achieve. It's sabotage, Major.'

'Are you sure?'

'I am quite sure the yield will be reduced.' Axelrod's voice became low and urgent. 'By how much, we won't know until we calculate the projected yield with an inert casing. I want to get some time on our electronic computer to work out how much.'

'How powerful does it have to be?'

'That's the point. General Secretary Khrushchev has ordered a one-hundred-megaton device. A hundred million tons of TNT equivalent. Yes. Roughly five thousand times more powerful than the Hiroshima bomb.'

'Fuck.'

It wasn't often that Vasin was moved to swear. Axelrod glanced up with faint distaste.

'That's one way to put it. A hundred megatons was meant to put the Americans in their place. Before Adamov began his alterations.'

'Will this device be bigger than the Americans' bombs?'

'The biggest hydrogen bomb they have ever detonated was about fifteen megatons, but that was by accident. Anyway, the Yankees got scared and started making them smaller.'

'Scared of *what*?'

Axelrod shrugged.

'I thought it was the *kontora*'s business to tell us what the Americans are thinking.' He ventured a quick, nervous smile before continuing. 'They got scared of the unpredictable effects, I suppose.'

'What unpredictable effects?'

'You wouldn't understand.'

Vasin ignored the insult.

'You're saying that maybe *Adamov* got scared?'

'Soviet scientists fear nothing, Comrade. We fear only fear itself. As the Yankees' President said once.'

'But isn't it dangerous to increase the scale so radically?'

'Thermonuclear bombs *are* quite dangerous, Major.'

'You don't think that caution in this matter . . .'

'Cowardice in this matter is a betrayal of the trust placed in us by the Party, Comrade. And deliberate sabotage would be, of course, *treason*.'

Axelrod spoke with a bravado that was just short of absolute earnest. Ah, thought Vasin. A weak man, flinging words stronger than himself in the hope that someone will believe him. He'd seen the type.

'And who knows about these changes?'

'Everyone. That's the way we have always worked here. A basic principle of our craft. At the beginning, the *kontora*, your chief, Beria, tried to impose compartmentalization of information. The Beard – Igor Kurchatov, the builder of our first Soviet device – refused. He insisted that here inside the program we have complete freedom to exchange information. And so it has remained. Therefore, Adamov handed out the new general design today to the heads of all the laboratories. Every senior academic worker has access to his department's copy.'

'And who knows about the exact implications of Adamov's change of design?'

'Protopopov, the chief metallurgical engineer. He'll be kept busy casting the new lead tamper. The rest of the design is pretty much unchanged.'

'Does anyone share your view that Adamov's plan is sabotage?'

'Every man here owes his advancement to Adamov. They are all his acolytes. For them the Director is the brain of the Institute. He taught them. He made them. They would never question his wisdom and judgment.'

'Except you.'

'Except me. But that is only because Fedya – Dr Petrov – took me into his confidence over their dispute. Adamov had been worried about the uranium tamper for months. They argued about it. Constantly.'

'Why didn't Adamov just overrule Petrov? If he is the law here.'

Axelrod flicked Vasin a helpless, pleading glance, as though he hoped to find an ally.

'Can I trust you, Vasin?'

Only a desperate man would ask such a question of his interrogator. Yet Vasin had heard it often enough. There comes a point when every man needs a confidant, even if his words will damn him.

'You can trust me, Axelrod. I promise.' Vasin even truly meant it.

'Petrov has powerful supporters. Had. His father, for instance. Petrov was the only scientist in Arzamas who could really stand up to contradict Adamov. And now he is dead. Adamov can do as he pleases.'

'Which implies, Adamov eliminated him in order to put his sabotage plan into action? Is that what you are telling me, Axelrod?'

'You answer your own questions so succinctly, Comrade.'

Vasin's own words to Adamov on their first day, flung back at him. But there was no longer any hint of humor in Axelrod's pale, agitated face.

—

The street was slick with mushy snow, churned into the consistency of porridge by pedestrians' feet. Vasin passed a team of street cleaners in padded coats who were scraping the sidewalk clean with broad steel spades. On the corner a snow-clearing machine, the kind they called a capitalist because of its rotating, grasping arms that scooped the snow onto a conveyor, idled waiting for the team to finish throwing the slush into the road. A comforting ritual of Soviet urban life.

Our death-breeding city, Maria had called it. Vasin began to see the menace. The hurrying schoolchildren, the laden shoppers, the comforting clack of closing tram doors, all formed a familiar screen of normality. But sitting on that tram, driving in that car, eating in that restaurant, were men and women who held the death of the world in their minds and hands. Everyone around him had his or her little piece of Armageddon to build. And somewhere beneath his feet, as he walked the corridors of the Institute, the bomb itself was growing in its cellar-womb like a monstrous baby.

Could Adamov have murdered Petrov because he wanted to be free to sabotage the bomb? The project that would crown Adamov's career with a mushroom cloud of such proportions that the world would stand awed? Why would Adamov wish to reverse a decade's progress, the research of thousands of his students and colleagues? Sabotage?

Or fear?

Fear of the 'unpredictable effects' of a device that was too powerful?

Vasin thought of Professor Adamov's pale, papery skin, the thin gray bristle that covered his scalp. What was inside that brilliant brain? To hold death, so much death, in that mind of his. And still be calm. How could one carry such knowledge and remain sane?

Vasin could find no obvious impossibility in Axelrod's story. The timing of the radical new design was certainly suspicious. But why would Axelrod seek him out? Why would a member of the Citadel, that ivory tower so tightly closed to the prying eyes of the *kontora*, choose a stranger from Moscow to confide in? Unless Axelrod was trying to deflect suspicion from himself?

Could it be, Vasin thought, that it was Axelrod who had slipped a hook into his mouth rather than the other way around?

V

Vasin found Efremov in his office at the *kontora*, sitting bolt upright at his orderly desk reading a report. Seeing Vasin's head poking around the door, he set the papers facedown and folded his hands into a steeple.

'Not disturbing you, Efremov?'

'Come in, Major.'

Vasin closed the door behind him. Efremov did not invite him to sit.

'I need a favor from you. An important favor. And I promise it won't involve distracting the golden minds of Arzamas.'

'You seem to have taken a lesson from the General this morning. I understand he set out some checks to your ambition and carelessness.'

'I took careful note.'

'And what is this harmless favor?'

'I need a copy of the lab report, specifically the thallium records.'

Efremov's face twitched involuntarily. He unfolded his hands and leaned back in his chair.

'May I ask why?'

'Because I have reason to believe they are inaccurate.'

The adjutant's voice became icy.

'I assume your informant in the Institute, Dr Vladimir Axelrod, has been making suppositions?'

'It's irrelevant who I talked to.'

'Oh, but it is relevant, my dear Vasin. Dr Axelrod is a known troublemaker. A man of dangerous views, an irregular personal life, and unreliable politics. And based solely on the word of this man, you wish to undo the work of an entire State Security team who combed through the laboratory records for days, cross-checking under supervision. To what end this folly? Are you driven to turn the Petrov tragedy into a detective mystery? Or did Orlov send you to deliberately disrupt the most important show of Soviet might in history?'

'Is that a no?'

Efremov struggled to keep his composure.

'It is a no, Vasin.'

Unbidden, Vasin squared a chair directly in front of Efremov and sat.

'You are covering for somebody. I don't know why, or for whom, but you're hiding something.'

Efremov smiled thinly.

'Wild accusations are all you have left. You seem to have me confused with some hapless patsy you have pulled off the streets of Moscow.'

'Come on, Efremov. You know this case stinks. And one of the reasons it stinks is your distinct effort to keep me away from it. Now you're going to tell me that the great project cannot be disrupted with days to go before the test. Doesn't a murderer who kills top scientists constitute a threat to your precious Arzamas? You saw what happened to Petrov's body. And you told me, "Maybe he deserved it."'

Efremov sat motionless, his palms on his thighs. A muscle in his jaw pulsed, but he said nothing for a full minute.

'What gives you the right to place your curiosity above the highest considerations of State Security?'

'You know, Efremov, you're not the first person in Arzamas to ask me that.'

'You have the arrogance of a fanatic, Comrade Major.'

'I have my orders and I follow them.'

'Vasin, you asked for my help. The best help I can give you is a word of advice. If you persist in your disruptive behavior, there will be negative consequences for you, personally.'

'Negative consequences from whom?'

'From patriotic men.'

'You mean General Zaitsev.'

Efremov's face had become stone.

'Very well, Efremov. I take careful note of your help.'

Vasin stood abruptly and made for the door.

'Comrade, all we want is for you to get home safely to your wife and child.'

Vasin froze and turned back.

'The work of a KGB officer is not always easy on a marriage. We do hope it works out for the best between you.'

Vasin scanned Efremov's face for signs that the man knew of the fatal knowledge that was contained in the transcript of his telephone call with Vera. 'Katya Orlova' – it was a common enough name. Had Efremov put two and two together and connected her to General Orlov? Vera could not have put him in more jeopardy if she was trying. Was the man needling or stabbing? Judging from the smirk on his face, needling. Or so Vasin fervently hoped.

He composed his face.

'Thanks for your concern, Efremov. I'll see her when my investigation is finished. And I mean *completed*.'

Vasin turned to open the door.

'One moment, Comrade. There is something I can do for you.'

'What's that?'

Efremov had risen.

'You wanted to see Petrov's apartment?'

Vasin hesitated. Something knowing and menacing had entered Efremov's tightly wound face.

'Why the sudden change of heart?'

'A gesture of goodwill, Comrade. We're on the same team, after all. Come. We can go right now.'

VI

Fyodor Petrov's apartment building was at the edge of town, a kilometer and a half from Lenin Square. It was a modern five-story building identical to Kuznetsov's, but facing a woodland park. The apartments had balconies, too, and a fenced-off children's playground, small signifiers of great privilege. Parked in the courtyard were a pair of official cars and two large Kamaz trucks. A policeman standing guard at the stairwell saluted Efremov and rolled his eyes to say: They're upstairs.

In the lobby the tiny concierge's lodge, decorated with magazine cuttings of Red Army hockey stars, was empty, and the doors to the ground-floor apartments stood ajar. Vasin glanced inside. Net curtains billowed on a breeze blowing through wide-open windows. A suitcase yawned, half-packed. The residents had left in a hurry.

Voices drifted down the stairwell. Vasin and Efremov ascended to the third floor, where a trio of green-uniformed backs huddled in conference.

'Comrade Officers?'

A narrow face framed in a lieutenant's collar bars turned irritably, then snapped to attention.

'Sir?'

'This is Major Alexander Vasin from Moscow. He wishes to see the deceased's apartment.'

The KGB subalterns exchanged glances.

With exaggerated formality, Efremov extended a cardboard box he had brought with him from the car.

'Take this protective gear, Major. The place is covered in . . .' Efremov searched for a more delicate word and decided against it. 'Radioactive puke.'

The box contained a gas mask of Great Patriotic War vintage. Vasin thought of the morgue – the breathing apparatus, the canvas suits, the hosing down.

'Are no other precautions necessary?'

Efremov gave a snort that said, probably.
'Take your time.'
'Thank you, Comrade.'
He pulled on the dusty mask.

Vasin had seen blood before. A schizophrenic electrician who had butchered his family in an apartment on Taganka. A drunk who'd been thrown out of a bar at Kursky Station and had returned with a pair of axes. But this crime scene was like nothing he had ever witnessed. The whole place, sleek bookshelves, modern furniture, shoes standing neat as parading soldiers, was covered in scarlet splashes. Great, extravagant puddles of bloody vomit, more than Vasin had thought one man could ever produce.

In the bedroom was a team of three technicians in the same hazard suits he'd seen at the morgue. They were stripping blood-soaked sheets from the bed with tongs and stuffing them into metal-lined plastic bins. The men acknowledged Vasin's presence with a brief glance, then continued their work. Vasin began to understand why Efremov wanted him to see it. The official investigation was over, so every scrap of evidence in Petrov's apartment was being systematically thrown away. On the bedside table a Geiger counter crackled steadily.

Radiation.

When Vasin first heard the word during a childhood X-ray, the nurse had explained that this magic ray was a force of nature that had been tamed by Soviet science. But here in Petrov's apartment the magic had been turned loose. Odorless, tasteless, and deadly.

Instinctively, Vasin tucked his bare hands under his armpits and looked about the bedroom from the doorway. On one wall hung a framed poster for a foreign film. Vasin squinted through the misted eyepieces of the gas mask. Gérard Philipe in *Le Rouge et le Noir*. Vasin backed out of the doorway and turned to the sitting room. Framed photographs of family holidays: Fyodor the golden boy, floppy-haired, posing with his parents on beaches and boats. Two

Doctors Petrov, father and son, standing side by side, both wearing lab coats. The older man sported a crumpled grin, while his son stared at the camera proud and invincible. Above the television hung a series of framed sketch portraits of Petrov, posing shirtless, executed in charcoal by a talented amateur. Vasin noted the signature: 'With all my love, V.' He hoped that Zaitsev's clowns had photographed the drawing. Within days, he guessed, this and every other object in the place would be buried at the bottom of a mine shaft.

Two of the technicians were starting to maneuver a laden bin into the corridor. One of them waved to Vasin and made a jabbing motion toward his wrist. Vasin stood uncomprehending for a moment then understood: time. Radiation exposure.

Back on the downstairs landing, Vasin tore off the musty gas mask and breathed deep. As he snatched it off he felt a dry rattle in the filter. Snapping off the cover, he found that the mask's snout contained nothing more than crumbled, useless wadding.

Efremov and his colleagues had gone. He could hear their guffawing banter in the courtyard. Seeing Vasin in the window, they tossed away their cigarettes and mounted their jeep. By the time Vasin had reached the front door, both they and Efremov had driven off, leaving him to find his own way to KGB headquarters. A pair of *kontora* goons sat in a Volga sedan, watching him pass like bored guard dogs.

Right now, he needed to get away from the poisonous air of Petrov's apartment. The visit had told him nothing, and in the process he'd inhaled God knows what dose of radiation. Or maybe not quite nothing. 'With all my love, V.' Vladimir Axelrod? The rain had thickened into a steady drizzle, but nonetheless Vasin struck out into the park in front of Petrov's building. Like most such green spaces on the edges of Soviet cities, it wasn't actually a park at all but a patch of primeval forest, abruptly demarcated by a line of asphalt that signified civilization's furthest advance.

Within a hundred meters, Vasin could have been anywhere in deep Russia. The birches and firs embraced him with the sweet smell of autumnal decay. Rain hissed on the yellowing branches.

He followed a lightly trodden path to a natural clearing where traces of ice had formed on the edges of a pond filled with black water. A bristling clump of marsh alders stood, their mongrel-brown trunks tangled fantastically. Vasin pressed on. The forest damp was like a balm against everything man-made and unclean. Behind him he heard the snap of twigs as his tails from the *kontora* struggled irritably through the wet undergrowth.

Abruptly the woodland ended in what looked like a wide fire-break. But, as he stepped out from the bushes, Vasin saw it marked the city's border. A tall barbed-wire fence ran along the center of a clear-cut area, perhaps a hundred meters wide, that extended as far as he could see in either direction. A gravel road ran alongside the fence, with guard towers every two hundred meters. Left and right, Vasin saw glinting pairs of glass lenses scanning him from the towers' walkways, the binoculars held by armed soldiers in rain capes. Opposite, through two lines of fence, the forest continued dark and impenetrable.

Vasin had reached the edge of planet Arzamas.

VII

As Vasin walked home, the sun dropped below the clouds and for a few minutes bathed the snowy streets in harsh, oblique sunshine. The tarmac shone with a lingering scarlet glow. The sight of the post office, its sandstone facade a rich gold in the evening light, held him for a moment. But after counting through the possible outcomes of a phone call home like a pauper auditing pennies, he could think of no good one. Even his final play, the one he promised himself he would never make, the one of total surrender and self-abasement, had not worked. He had thrown himself on Vera's love and mercy but found that there was none left over for him.

There had been a time, once, when Vasin would confide in Vera. Sharing his indignations, his small triumphs, talking over the mysteries of his police-work cases. Never much of a reader, Vera, but she had liked Sherlock Holmes as a girl. Her breathlessly inventive explanations for his homicides reminded him of a younger version of himself. He could never quite bring himself to tell her that Moscow's murderers were, for the most part, a depressingly predictable bunch. Weeping drunks, beaten wives, the furious human dramas of the communal apartment kitchen where a missing chicken could lead to bloodshed. And then there were the occasional turf wars between such underworld figures hardy enough to exist in a city as saturated with police and vigilant citizenry as Moscow. A gambling club shoot-out. A prostitute with her throat cut. No complex motives, nothing that Holmes and Watson would ever feel the need to light a pipe to ponder, ever appeared in the files that landed on Vasin's desk. No speckled band snakes, no phosphorescent hounds. Only thieves' pathetic ideas of honor, profit, and survival. The desperate things human beings with no options left did to each other.

He and Vera had met, in approved fashion, at a Young Communists' dance. A dusty hall in late summer, the smell of sweet teenage sweat and floor polish. A loud emcee with hair oil running down his temple, a band made up of pimply young men in square suits. From among a crowd of her girlfriends, gripping bottles of soda like protective talismans, Vera looked out into the hostile territory of menfolk with an expression of calm and tolerant appraisal, strangely without ambition. Vasin had of course been too nervous to ask her to dance. She did it instead, accompanied by a girlfriend. She spoke for both of them.

'Will you sit there all evening? The music will finish soon.'

There was annoyance in her voice, an accusation that the boys weren't doing their bit. As he and Vera danced, awkwardly, hands on hip and shoulder and a chaste arm's length apart, Vasin felt surprise and relief. So she knows how this all works. Outside, after the dance, they sat side by side on a bench among the long shadows of a small park. Again, Vera had taken him in hand.

'Well?' she asked. Her tone was that of Vasin's mother, drawing

her boy's attention to some undone chore or broken promise. 'Will
we kiss?'

She turned her head, eyes closed, mouth half-open, waiting. Her
mouth tasted of sugar syrup. When his hand closed on her small
breast she slapped it away.

'What are you doing? It is not seemly.'

Vera always knew what was expected. Her life proceeded according to a timetable of convention known to everybody except Vasin.
On their third date, to see a comedy film at the Arbat Cinema, she
had allowed him to put his hand up her skirt. He felt only hot, taut
nylon, and an impenetrable fortress of underclothes. It was understood that a proposal was required to get further. They were both
nearly twenty. It was time.

Vera's eyes were cool and gray, and she had a squat Russian home-
liness he found comforting. She was accepting and grave. When he
suggested they get registered, after queuing for an hour for a table
at the Aragvi restaurant and working their way through a bad
Georgian dinner, she did not smile exactly, but her mouth appeared
to relax.

'*Nu na konets*,' she said. 'Finally.'

Vasin assumed that meant yes.

Vera had embarked on the arrangements for the wedding sto-
ically, keeping him updated on the process of obtaining champagne
coupons and a dressmaker's appointment as though he had laid a
burden upon her. She allowed him to make love to her, on a friend's
sofa. The event was signaled long in advance. Arrangements had
been made, keys passed over with a knowing look. Vasin, in his
nervousness, had drunk too much. Afterward he remembered only
Vera's sigh when it was over, far too quickly. She had allowed her
man to indulge his bestial instincts.

Vera's mother, Margarita Ivanovna, lived in a new five-story
building on the outskirts of the city. She was not yet forty, but her
young face was worn out and framed by hair gone dead. Husband
lost in the war, officially. Unofficially, Vera had confided, run off
with another woman. Margarita Ivanovna had eyed the flowers that

Vasin brought disapprovingly, as though the gift might carry some sort of obligation. The bouquet was evidently too gaudy for her. What was this young man trying to cover up? In time Vasin learned that too modest a gift would also have been wrong for his mother-in-law. But by then he had grown used to his role. As a man among women, he stood for all that was wrong and unjust in the world.

Vasin and Vera received the portion of happiness that they had been raised to expect. As the son of a war hero, Vasin had to wait only three months for a one-room apartment in a concrete high-rise on Lenin Prospekt. Vera had been assiduous in finding correct furniture, buying plastic flowers, hanging Hungarian curtains. She even learned to enjoy their sessions in bed, or at least pretended to. Night after night, she had worked her way through the recipes in *1001 Things a Good Housewife Should Know:* borscht, cutlets, beetroot salad, fish soup. The training manual for a good Soviet wife, the points to be ticked off one by one on the road to domestic perfection. In their tiny kitchen, over cups of tea, she had listened to her young husband speak of the trials of the police academy, then the force, then his murder cases, or at least sanitized versions of them. She had found a job in the accounts department at the nearby Dom Tkani, Moscow's biggest draper's emporium, and settled into the petty politics and jealousies of the office with enthusiasm. She had smiled, at least once, almost every night.

Nikita had been born two years after their marriage. Vasin had first seen the infant as a swaddled parcel, displayed from behind a sealed third-floor window of the House of Births on Nikitskiye Vorota as though to prove that Vera had not been malingering. No men were allowed into this temple of female suffering, for which the husbands gathered on the sidewalk outside were clearly held responsible. The birth had been a bad one, Vera had told him, and he'd pretended to know what she meant. Never put me through that again, she had said, as though the whole thing had been his idea. He did not protest.

At first Vasin had been grateful that the squalling, red-faced creature occupied so much of Vera's attention. He had been displaced in her life and affections, and was relieved. He discovered from his cop colleagues that real men were expected to drink. Soon, he began to enjoy the confident glow that booze gave him. At a certain point, his swaying gait could almost pass for macho swagger. At a Women's Day party organized for the officers of the Moscow Criminal Investigation Department, Vasin had found himself drunkenly propositioning a pretty blonde, the sister of a colleague's wife. To his intense surprise the girl had taken him by the hand, as though weighing his wedding ring, then turned and led him out of the dance hall and to his brother officer's apartment. They had screwed with desperate haste on the sofa, right in front of a toddler who stood and watched them wide-eyed from his crib. As they lay together afterward, breathing each other's sweat, they had listened for the whir of the lift that would announce the return of the child's parents. As they made love the girl had moaned and writhed in his arms like a wild thing. That night, Vasin realized, he had discovered lust. Sex was not just an unclean bodily need to be discharged but could be an ungovernable hunger. He had never before felt lost, weightless, outside time. He imagined he could taste the girl's salty kiss on his lips for weeks afterward. He wrote to her at her women's dormitory at a textile factory in Ivanovo, but she never wrote back.

In the banter of the detectives' operations room, they teased Vasin as a *babnik*, a ladies' man. The truth was that he didn't dare. He was handsome, or so the teenage secretaries and shopgirls and waitresses and even, once, a pretty trolleybus conductress who chatted to him with bright, inviting smiles, told him. Neatly dressed, closely shaved, hair brushed. A model Young Communist, a young man going places. But women, for Vasin, remained fundamentally unknowable and dangerous. Every winning female smile seemed to him to conceal a terrifying hinterland of hysteria, pregnancy, and disaster.

When had he and Vera stopped talking? Parenthood had tipped over the rhythm of their lives in all the usual superficial ways. But in fundamental ways too. And it had never righted itself. Vasin remembered a line from a Mayakovsky poem: the boat of love smashed on the shore of daily life.

It was not as though Vera was unhappy, or at least, no more discontented than she had a right to be, than anyone else seemed to be. Vasin had concluded early on that unhappiness was inevitable in any human relationship – unhappiness suffered and unhappiness inflicted. He had never taken it personally. It was, evidently, a law of nature. The sadistic inclinations of Vera's supervisor at work, the lack of emulsion paint in the hardware shops, the crowds on the rush-hour trolleybuses, all these blended with her discontent with his own personal slovenliness, the child's laziness and sleeplessness, the neighbors' noisy radio. All these were simply the rhythms of life. Vera's slow-burning eyes, nightly censuring male insufficiency across the dining table, were things simply to be borne. Just as his colleagues' philistine disdain for culture was a thing to be borne, his boss's contempt for rules and procedures, some suspects' stubborn refusal to accept that fate had ordained them to be victims. Lives were imperfect; wives were imperfect. How foolish his unmarried colleagues were to be afraid of loneliness.

But there was something deeper. Vera had no interest in changing what was around her. Worse, she had no concept that the world could be changed.

Vasin did.

That stubborn streak of righteousness he had inherited from his mother. Not rebelliousness, but rather its opposite, a dogged insistence at taking the system at its word. 'Citizen, we have the right!' had always been his mother's favorite phrase. Vera had no such faith. She trudged irritably through a bleak world where hardship could be overcome only by moments of luck and acts of low cunning.

Yes, Vasin realized with devastating clarity that his wife was cynical, and that she was stupid. He had felt a nerve in his head tighten. His usual misery had uncoiled with its inevitable routine. 'You are naïve,' she would sneer. 'You would rather be right than get ahead.' Vasin longed to retaliate, to fling back his own contempt. But he knew that this would mark a defeat for him, that any reconciliation would be only on her terms. And this he would not allow. Because he knew that he was right. Or at least, that right existed, somewhere, if he could only find it.

Naïve? Perhaps. Vera's words had stung because they were true. Every day of his career as a policeman he saw his colleagues framing witnesses, pinning spare crimes that had been bouncing around the books onto hapless suspects, driving practiced fists into soft bellies, scrambling at the end of every quarter to round up some likely patsies to fill the crime-statistics quotas. He knew all this, and knew it to be wrong, a travesty of Soviet justice.

But a weasel thought had sneaked into his mind.

What if crookedness was the way the system was *meant* to work? They pretend to pay us, we pretend to work, his colleagues were always joking. What if his job were to pretend to protect the people, to act a bit part in a giant parody of police work?

Vasin had never thought to call such thoughts political until his friend Arvo did. Arvo Janovich Laar was an Estonian, but the son of a Bolshevik hero, so it was all right. Laar was the uncomplaining butt of constant joshing about his countrymen's supposed timidity and phlegm. He had a singsong Baltic accent that their fellow detectives never tired of imitating.

'Hey, Laar! Heard the latest one? Russian gets on a bus in Estonia. Asks if it's far to Tallinn. Driver says, "No-sir. Not-far-at-all." They drive for an hour. The Russian guy asks again, "Is it far to Tallinn?" Pig-fucker driver answers, "It is-now."' Uproarious, locker-room laughter. Even Arvo would join in. But he knew, and they all knew, that what the laughter really said was that Balts were sneaky, defeated, Russia-hating little assholes.

The bullying drew them together, though Vasin knew it and didn't like it. Their outcasts' friendship had been assigned to them by the group, like the roster of a detective team. Vasin and Laar, the weirdos. But Arvo was easy to like, counter-stereotypically cheerful and essentially optimistic. Vasin recognized something of Arvo in himself. His refusal to bend before the crowd. A core of pride.

They found themselves alone together one afternoon in an office that smelled of stale tobacco smoke and disinfectant, spilled booze and vomit, the after-murmurs of a recent party that they had missed. Gingerly, as though unwrapping contraband, Arvo ventured a joke of his own.

'An Estonian goes to the polling place, prepared to vote. He is handed an envelope and told to put it in the ballot box. But instead of following instructions, he starts to open the envelope. "What are you doing?" yells the woman. "I just wanted to see who I was voting for," replies the Estonian. "You imbecile! Don't you know this is a secret ballot?"'

Vasin had quickly hoisted a smile, but Laar spotted it for a false flag.

'What? Not funny?'

'It's funny, Arvo.'

Laar had looked at him for a long moment, opening his mouth just wide enough to let the point of his tongue caress his upper lip. Then he closed it and allowed a further pause for uncomfortable thoughts.

'You're a believer, Vasin.'

'In God?'

'No! For heaven's sake. I mean . . . a believer in . . . the order of things. Don't get me wrong. I'm glad someone is. I admire you for it. You believe that things should be as they *should* be, not as they are. That things can be done properly. Will be done properly, one day. You can't laugh at the system because you *believe* in it.'

Vasin set down his paper coffee cup and thought, Yes, I do.

It had never properly occurred to him that there were people, respectable people, who did *not* believe.

'You think our bosses believe in anything apart from collecting kickbacks from the Chechens and the Armenians and the Tambov boys?'

Vasin was shocked. Not that he hadn't heard the rumors about their boss. Of course he had. But to hear his friend discuss them so casually seemed a sacrilege against the gravity of the accusations.

'You're saying that it's true? Have you reported it?'

153

Laar sat back with a sigh.

'And what would be the point of that? To do my duty as a citizen, and get myself reassigned to Syktyvkar?'

Laar drained his cup of thin police coffee before continuing.

'You know what I can never understand about you Russians? You love anarchy, each one of you. I know you fuckers. You all have rebel souls. Every one of you wants to screw the system. But together, you have a terror of chaos. You'll go to any lengths to prevent it. I always wondered why. Perhaps because you know yourselves too well. You know what you'll do if suddenly nobody is watching over you with a stick. That's my profound thought for the day.'

Vasin thought about that term, *rebel souls*, for days, in the queue at the bread store, in the line at the cafeteria, in the scrum of commuters as they positioned themselves to fight their way onto the Number 31 trolleybus home. Rebels? These?

But at the same time, what were these cancerous secret doubts that were eating away at Vasin's righteous heart? He searched inside himself for a cool-headed voice that would explain the gap between the sordid daily necessity and the far, shining vision. But as the days went by Vasin began to feel the vision slipping too far away to be grasped, into the realm of myth. He remained convinced that belief was the only gravity that could hold his contradictory world of duty and lies together. But he was no longer sure he had preserved enough of it to make the center hold.

Two years after his conversation with Arvo and newly recruited to the KGB, he'd met Katya Orlova. It felt like justice. 'Frustrated salaryman seeks his nemesis, sex-starved Juno preferred. Gods' wives only.' Vasin had embarked on the affair in the certain knowledge that there was no possible universe in which it could have a good ending. He felt like an alcoholic reaching for the bottle he knew would kill him. As Katya's monumental breasts smothered him in bed, Vasin experienced for the first time the abandon of self-destruction. It was like a coming of age: After this, there could be no redemption. Because he had finally become a faithless man, and deserved none.

Exactly how Vera had found out, he didn't know. The bitches' coven of the *kontora* wives had woven a sticky web of rumor that had spun quickly across the Moscow phone system to reach Vera's ears. As he knew it would be, this was Vasin's grand debacle. The moment of weakness that he sensed Vera had been waiting for all her life. The moment that she and her mother had secretly known would come: all that was rotten and duplicitous about Vasin had finally been confirmed. At last! We knew all along what you were. Finally, his behavior had met their low expectations. And of course, what Vasin hated most was that they were right. His own weakness, his pathetic lust and lack of self-control, had given this crowing chorus of frustrated women the justification they needed to pull his life apart. A just fate had come for him, even if it never seemed to come for anybody else. And it came in the form of his wife, transformed by his own folly from a lonely, silly woman into a righteous avenger.

How ironic that Vera's revenge should break over his head while he was here, in Arzamas, the assignment that was testing his faith in himself as an agent of justice to the breaking point. Efremov was right. Vasin did think that he knew better, that his precious right to get to the truth trumped everything else. And Maria was right. He'd spared her from the asylum, then come to her to present the bill. And Vasin knew that he was not a righteous man. Thanks to Vera, all Arzamas might soon know it too.

CHAPTER SEVEN

THURSDAY, 26 OCTOBER 1961

FOUR DAYS BEFORE THE TEST

I

When Vasin woke, the morning light was already streaming through the orange nylon curtains. Yesterday's sleet and snow showers were a memory. A bold autumn sun glittered on rows of brand-new apartment windows, shiny as mirrors. From somewhere down the street came the babble of a radio. Vasin was alone in the apartment. Kuznetsov had left some buckwheat porridge on the stove, fruit compote in the refrigerator.

In the bathroom the freshly painted pipes shivered slightly when he turned the tap. The showerhead dangled a tantalizing thread of scalding hot water, then abruptly burst into full steaming force. Arzamas water had a peaty aroma to it. Vasin scrubbed himself, for the fourth or fifth time since he had returned home the previous evening, praying that the invisible contagion of Petrov's apartment had not entered him. Once again he cursed Efremov. The dud gas mask had been a terrifying warning about how his life could be effortlessly destroyed by the secret poisons of Arzamas. And again he replayed Efremov's words about Vera in his head, their tone, the cast of Efremov's narrow face, his contempt for Orlov and his ignorance of the hidden power that Special Cases wielded. Even if the local *kontora* forbore to take his life, did they possess the knowledge that could destroy his career?

Vasin was scared. Frankly, scared.

In any other circumstances, he would have taken Zaitsev's advice and cleared out of town. Not even his own curiosity about what the hell they were covering up would have held him here. Efremov might well guess the truth about Katya Orlova. In any case Vera herself would blurt it to someone else soon enough. Vasin was trapped in this accursed city. A bomb of his own making was ticking under his life. Staying on was a risk. But cracking the Petrov case was also his only lifeline. Orlov, it was clear, had sent him to Arzamas because the old man's instinct had smelled the possibility of netting a golden fish. And finding that fish was Vasin's only chance to ever protect himself from the wrath of Orlov that he felt brewing on the future's horizon like a gathering storm.

Vasin repeated the dry-cleaning routine he'd used the previous day to shake off his watchers. Once again he was lucky, a bread truck this time, beeping its horn impatiently outside the service gate of the *kontora* headquarters as he slipped out of the back door. Not a circumstance to be relied upon, and a violation of the basic rule of countersurveillance. People notice patterns. A stranger wandering through the courtyard will be ignored. See him twice, and you remember you've seen him before. Same went for the stationmaster, wide-eyed with nervousness when Vasin appeared in his office a second morning in a row. Vasin mouthed a silent prayer to the God he didn't believe in that Orlov would have something for him.

But the line to Gorky didn't connect.

Vasin dialed the four-digit code again and again, only to hear a continuing dial tone. He tried the time-honored Soviet technological fix of first resort, banging the receiver on the side of the heavy steel box. For good measure he tried the second resort, too, which was swearing in an angry whisper.

By the time the line finally connected it was a quarter past ten. Fifteen minutes late.

'The Comrade Chief Engineer is in a meeting.' Orlov's secretary delivered the bad news with the usual tart satisfaction. 'You'll have to wait till the next train.'

So funny.

'I am afraid there will be no next train. Can you tell the Comrade Engineer that there is an emergency? Eight locomotives are about to collide.'

'Wait.'

Vasin could imagine the secretary's sour moue as she took her time sauntering over to the conference room.

At length Orlov came on the line.

'*Eight* locomotives?'

'Sir.'

'Well. We must take steps to avoid such an eventuality. Interesting news. On your request.'

The background cacophony on the line suddenly broke into their conversation. Two husky male voices began speaking over Orlov.

'. . . and I said to that cow, I'm sick of the fucking sight of you. Every day I come home and there you are, sitting on the sofa like a sack of potatoes. And she says, If you'd stop drinking that crap you might appreciate me more. And I said, I drink crap for economy's sake, woman! Save money for your extravagances . . .'

'*Fuck off the service line now!*' Orlov's voice was at its most commanding. '*Bosses* are talking.'

'Okay, okay! Calm down, bitches!'

The interrupting voices went silent with a click, leaving only the usual background of low chatter.

'Your sharp nose has not failed you. I have had some responses from the archives. About the derailment back in 'thirty-seven.'

A pause as Orlov spoke aside. Clearing the room, Vasin guessed.

'It seems that Party A, repeat, Party A, was punished for his role in the accident after an anonymous denunciation. From a colleague at the same depot. The Kharkov depot. This source appears in the files under the name Kukushka. The Cuckoo. Yes. And further inquiries have established that the true identity of this source was indeed Party P. Repeat, Party P. Who was at the time Party A's immediate subordinate at the depot. Repeat back to me what I have told you.'

Orlov's words came over the antique phone line like an old gramophone record, a voice from the past.

'Party A was punished as a result of a denunciation from Party P, senior.'

'Exactly. There is more. After Party A's imprisonment, his wife divorced him.' Orlov was now evidently reading from a document. His voice droned matter-of-factly. 'She wrote a letter to the court accusing her husband of anti-Soviet activity. She saved herself. Spoke against him at the trial. Later married a . . . Yes. An Army officer.'

'This is remarkable information, sir.'

'Indeed. Indeed. Though not an entirely unusual situation. Given the, er, state of the railways during that period. But. It is clear that Party P has much forgiveness to ask of his old comrade.'

'How do you wish me to proceed, Comrade Chief Engineer?'

Abruptly, the other faint voices on the line went dead. A loud series of fast clicks replaced the hubbub, ominous in the sudden electronic quiet. Orlov, his voice almost drowned by the clatter, spoke loudly and emphatically.

'Proceed with your usual discretion. Be careful that nobody knows the target of your inquiries. You will take no action and you will report your findings only to me. Only to me. You will take steps to ensure that the local branch remains in ignorance. But find out everything. We must avoid any further derailments.'

'Yes, sir. But, General. May I ask, why am I really here?'

A sigh on the line like wind singing in the wires.

'Comrade, are you scared?'

'Yes, sir.'

'That means you have knowledge. Only a fool would not be, if what we suspect is true.'

In his sudden anguish Vasin abandoned all pretense of coded speech.

'They have threatened me, sir.'

'Who?'

'The local *kontora*. They're very serious. Sent me into a radiation zone.'

'No, you allowed yourself to be sent.'

'As you say, sir. But I just want to ask – is there anything you have not told me? About Arzamas? That I should know?'

Orlov's tone hardened.

'Take care that your fear does not make you foolish.'

'Sir, if an accident should occur, I want you to know that Major Efremov of State Security . . .'

'Major Efremov will doubtless be in charge of the investigation into your death. I will be in charge of recommending you for a posthumous decoration. And it will be one that does honor to your many talents. Your wife and child will be well taken care of by the State. You have my word on that.'

On the crackling line Orlov's chuckle sounded like the rustling of leaves.

II

From a phone box on Lenin Square, Vasin dialed the Adamovs' home number. He heard a telltale hiss on the line, even as the ringing continued. Someone was listening in.

Maria picked up on the fifth ring.

'Maria Vladimirovna Adamova? This is Major Alexander Vasin of State Security. My apologies for disturbing you at home.' At the other end of the line he heard Masha's small sigh. She had understood the meaning of his exaggeratedly formal tone at once.

'It's no trouble, Comrade Major.'

'I am afraid that I have some further questions. May we meet at the same place we conducted our last interview in half an hour?'

'I will be there.'

Vasin listened for the clicks as both Masha and their silent listener replaced their receivers, then swore softly.

No point in hiding. In fact the opposite, to try to shake his shadows now would only alert them. They would be following

Maria too. The only choice was to hide, as Orlov had advised, in plain sight. He caught a tram to Lenin Park.

Fat, freezing raindrops cut through the dirty snow. Vasin watched Masha's blue raincoat approach down the boulevard, trailing a nondescript gray figure in her wake. The park's café was crowded with grandmothers and young children sheltering from the rain, so they walked instead to the empty bandstand. No one, to Vasin's knowledge, had so far succeeded in bugging a park.

'Hello, you.' Maria seemed unnaturally cheerful. She had made herself up, and under her mac she was wearing a smart woolen twinset. He saluted her, for the benefit of the watchers in the trees.

'Maria.'

'I can't talk like this.' Maria scanned the strollers who dawdled on the edges of the wide lawn. 'It's like being on a microscope slide, some great eye watching us from above.'

'Act naturally. There's nothing wrong with you and I meeting.'

'Meeting in a park? In the freezing rain? Are you in trouble?'

'Not yet.'

'Am I in trouble?'

'No. Neither of us is in trouble. They're curious about why I am talking to Axelrod in these dangerous times.'

'If you spoke to him, then you know. There's no way that man is a good liar,' she said. 'What I said about Axelrod and Petrov is true.'

'Perhaps. But Axelrod told me some things of his own. About Petrov and your husband. Explained that they might have had reasons to disagree. Serious reasons.' Vasin made sure to hold her gaze.

Maria took a step away from him, pulling her mackintosh tighter around herself and scanning the passersby.

'The little shit. What kinds of disagreements?'

'Believe me when I tell you that it's complicated.'

'You can't . . .' A note of fear sounded in Maria's voice. 'You can't really believe that Adamov had anything to do with Fyodor's death? You met him. You saw him. He couldn't hurt anybody.'

Vasin thought of the film he had watched his first night in Arzamas and the apocalypse soon to be unleashed by Adamov, tossing

the plane miles up in the sky, a firestorm racing across the land
below. Was this the creation of a man who would never hurt
anybody?

'Whatever Axelrod told you about Adamov's supposed argument
with Petrov is a lie. All his colleagues will confirm it.'

'Has Adamov ever spoken about his experience in the camps?'

'What's that got to do with anything?'

'Did he ever speak about who denounced him in 'thirty-seven?'

Maria set her jaw defiantly.

'What is this? What could you possibly know about the affairs of
great men, a little pawn like you?'

Anger sparked in her eyes. Perhaps desperation, too, a desperate
desire not to know. Despite himself, Vasin felt the sting of her
words, followed by a sudden, vicious impulse to hit back.

'Fyodor's father denounced your husband. The great Academician Petrov sent your husband to the Gulag.'

Vasin closed his eyes for a moment, shocked by his own words.
He had just blurted out the very thing that he should have kept
most secret. The professional in him knew that information should
be hoarded like dry gunpowder, then carefully laid at the weakest
point of your opponent's defenses for detonation at the most devastating moment. Instead he had ignited it in a single, useless whoosh.
For a moment he had the unsettling feeling of not recognizing himself, as though he had been momentarily possessed. But by what?
Why the sudden desire to lash out at Masha?

She stood very still, her eyes on the ground. Vasin could not
unsay what he had said. What was the phrase that Orlov had used,
once, of the momentum of a confession, showing off his wartime
German? Yes – the desperate man makes a *flucht nach vorn*, a flight
forward. An act of desperation, compelled by a previous false move.
Except now it was Vasin who was the foolish talker. He had no
choice but to plunge on.

'It's in the files. Fyodor's father destroyed Adamov's freedom, and
his family.'

When she finally spoke, Maria's voice came from a distant place.

'Adamov's first wife divorced him.'

'She repudiated him. Testified against him in court and denounced him as a traitor to the Motherland. *Then* she divorced him.'

'He never told me.'

Tears abruptly ran down her cheeks. Maria wiped them away, angrily, as though they had betrayed her.

Vasin felt a shameful glow. The sordid power this secret knowledge gave him. With a few scything words, her pride had been slashed down like grass. But after the glow came another urge, rising from an entirely unfamiliar place inside himself, to step forward and comfort her.

Her gaze, when it met his, seemed at first to signal a desperate vulnerability. Then like molten glass her eyes slowly hardened till they were brittle and unyielding.

'Why tell me such things? What is it that you want from me, Chekist?'

Her voice was bleak. Vasin had wanted to hurt her. Instead, it seemed, he had shattered her, and their fragile trust.

'If Adamov knew who betrayed him, it would provide every reason to kill Fyodor. Petrov took away his family, so he destroyed Petrov's son. You can see what it looks like to the *kontora*. I thought you needed to know.'

'You want to *protect* me?'

Vasin could find no words to answer her. But he knew why, and the sudden realization chilled him.

'Protect me from Adamov, or from your own *kontora*?'

'I need to find out the . . .'

'You're about to say "truth". Your favorite word. No need. Spare me your truth.'

Vasin began to reply, but she shook her head firmly.

'Vasin. Alexander. I will tell you this: In all our years of married life I have never known more of Adamov than the part he chooses to show me. It's as though his soul is wrapped in tightly wadded layers of silence. But I know this, if he was angry once, it was long ago. The past is the past. For both of us. Yes, he carries death inside himself every day, but he is not an evil man. He is not a violent man. He could not kill Petrov. And over some ancient grudge? Never.'

'Did he know about you and Fyodor?'

'Can you spare me your fucking detective story plots? Adamov is not some jealous French lover boy.'

'Maria, I believe you.'

'You believe me, *but*?'

'Axelrod. I need proof.'

She took Vasin's hand and squeezed it. After a long moment she seemed to come to some private decision.

The tram was packed with lunchtime shoppers. Maria's thin body pressed against Vasin's in the crowd. Two *kontora* Volgas followed them, overtaking the tram in relays. Doubtless another car would be waiting outside the Adamovs' apartment. Quite an entourage.

As she fumbled with the keys, Maria kept her eyes down, avoiding Vasin's. Once inside, she double-locked the door and tossed her coat on the floor. Vasin made a quacking gesture with his hand. Keep talking.

'Thank you for your good citizenship, Maria Vladimirovna,' he said loudly. 'Your remarks have been most helpful. Would that all our countrymen were so zealous in assisting the organs of law enforcement.'

He walked over to a telephone receiver that stood on a table in the hall. Silently he pointed at the apparatus and tugged his ear. Maria nodded in understanding and beckoned him to follow. She opened the door to Adamov's study, an austere room filled with neatly labeled files and stacks of journals, and pointed out the other receiver. Then she led him to the kitchen, where a third telephone hung on the wall.

'May I offer you some tea, Comrade Major? I feel a chill from our walk.'

'I felt a chill too. But only if I am not keeping you from your busy day?'

'It is I who fear that I may be keeping you from your important duties.'

'Not at all. Tea. Yes, please.'

The presence of invisible listeners made their conversation easier. Maria filled a kettle, struck a match, lit the gas.

'Excuse me for a moment, Comrade?'

She returned with a large brown envelope and placed it silently on the kitchen table in front of Vasin.

'Would you like some jam with your tea?'

Vasin slid out a smooth wad of black-and-white photographs. On the top was a typewritten card with a single line of text, the letters punched in angry capitals. It said:

HER OR ME?

The images had been developed and printed by an amateur in a home lab. The lighting of the photographs was pale and spectral, and from the little Vasin knew of such matters he concluded that the negative was made on fast film, for the print was also grainy. The first was a portrait of Petrov, shirtless, lying languidly on a divan. A half-smile on his handsome face, he seemed to be humoring some joke made by the photographer. In the next shot Petrov mugged for the camera, pouting. In the next he lay back, stately, remote, floppy-haired. A man who knows he is beautiful. The photographer had taken a step backward, revealing that Petrov wore no trousers.

'I made this jam myself. You must try some. I am not very domestic, really, but I do know how to make jam. The only thing I can cook.'

Masha's face was tight with tension. She stood opposite Vasin, fighting to keep her gaze off the photographs.

'Thank you, I would love to try some.'

Stiffly, she walked over to the cupboard to retrieve a jar, a saucer, and a spoon.

The next photograph was blurred, Petrov swinging his bare legs onto the floor, his genitals a smudge of black in the badly lit photo. Two more images of his naked torso, arms and legs in movement, as if he was dancing. In these photographs Petrov seemed completely self-absorbed, eyes closed, as though bathing in the loving regard of another.

'I found a splendid bush of sea buckthorn, hidden in a stand of birches. Somehow none of my neighbors have found it. I keep it secret from them. You see, everyone in this town has a secret. I can't tell you where it is or you might report it. Please, taste.'

Maria placed a saucer of the bright orange jam in front of Vasin, keeping her eyes averted from the table.

'It's most excellent, Maria Vladimirovna.' His voice felt unnaturally loud. 'Better than my mother's.'

Vasin turned over a second set of photographs, these smaller and printed on more sensitive photographic paper. They were harder to make out, more grainy, a confusion of blobs and rounded surfaces. Close-up shots of flesh. Intimate photographs of Petrov, wantonly lying in his bed, legs akimbo. Vasin recognized a corner of the *Le Rouge et le Noir* film poster he had seen on the bedroom wall in Petrov's apartment.

The final couple of photographs were badly aligned, apparently taken using a timer by a camera balanced at the foot of the bed. Two naked men lay twisted in each other's arms, their faces invisible as they shared a deep kiss. One was Petrov, his muscular shoulders and gelled hair clearly recognizable. The other figure had a skinnier frame, long hairless legs. Axelrod?

'Good God.' Vasin couldn't help himself. He looked up inquiringly at Maria, but she had turned away from him. The silent listeners sat among them like invisible guests at the table. 'I seem to have spilled some tea.'

She said nothing.

'Maria Vladimirovna.' Vasin stood. 'You said there was a household task you wished me to help with. Allow me to assist you and then I'll be getting along.'

'Thank you, Comrade Major. That would be most kind of

you.' Maria's face was taut with held-back emotion. Vasin gestured toward the bathroom. She swallowed and composed herself. Before she spoke she scooped hair from her eyes and arranged her face as though for a film take. 'Could you please help me move the clothes-washing machine? The drainage pipe has fallen down the back.'

'I'm happy to assist you, especially when the master of the house is engaged in such heroic work for the Motherland. Such domestic details must not be allowed to distract him.'

Vasin closed the bathroom door. Behind it stood a large steel tub covered in cream-colored enamel. An electric wringer was attached to one side of it, and an electric cable connected it to a socket. A stylized chrome decal on the front read WESTINGHOUSE in Latin letters.

'Like the appliances?' Masha's voice was bleak.

'We don't have much time.' Vasin turned on a tap and lowered his voice. 'Where did you find them?'

'Fedya had been avoiding me for days. I went around to his apartment and found *them* there. They'd been having a row and Axelrod was actually crying. When I showed up, he ran out of the apartment like some hysterical woman.

'I asked what had upset Axelrod so much, and Fedya said his friend had girlfriend problems. But the look Axelrod gave me as he left, it was pure hatred. And Fedya was in a state too. He put on a coat and ran down the stairs after him, and I'm left there wondering what the hell is going on. Fyodor, he was always so cool. So aloof. You know. Those steady eyes, his easy smile. Fyodor's voice, that elegant Moscow drawl, so full of confidence. His life had always been so carefree. There was no core of pain to him. Sometimes I thought Fyodor was like a visitor from some hot foreign place that had never known war or hunger. And suddenly, there he was in the middle of some hysterical scandal. Running out into the night. Made no sense.'

'You had no suspicions? How could you not have known?'

Maria shrugged with her whole body.

'Who cares, now? But no. It was good to be with him. He was a boy with the sun in him, my grandmother would have said. But then I found the photos.'

'How?'

'He must have stuffed the envelope down the back of the sofa cushions when I arrived. I wasn't searching. Okay, you don't believe me. I don't care. But I just noticed the package. Pulled it out. Read the note. And saw what you just saw.'

'And did you confront him?'

'I never saw him again. Not until the night he died, at our house. It would have made awkward dinner conversation.'

' "Her or me." Axelrod wanted him to leave you and be with him. Who did he choose?'

Maria looked up, puzzled.

'Meaning?'

'Who did Petrov choose, you or Axelrod?'

'What kind of a question is that? I'm not a can of fish on the shelf.'

'You say you never saw Fyodor after the apartment. Did he call you?'

'We had a private system to get messages to each other. What public phone box, what time. We spoke, one time.'

'And he apologized?'

Maria shuddered and shook her head.

'Fedya knew I had found the photos. He vowed he'd been seduced, he was a victim. The devil had gone inside of him. Fedya was brilliant and ambitious. He wanted to be an academician. A minister. He couldn't have someone like Axelrod around, black-mailing him. He told me he was planning to get Axelrod sent somewhere far away where he couldn't hurt us. Fedya had the con-nections to send him to a uranium-enriching plant in Siberia. He wanted to break him. Shut him up.'

'Did he tell Axelrod this?'

'I only know what he told me. I'm untouchable, he said. Axelrod is nothing. Fyodor had powerful friends who could bury any scandal. Along with Axelrod. All I know is a few days later Fedya was dead.'

Masha began to shake, gently, slurping back tears like hiccups. Abruptly she stood and embraced Vasin. The gesture was impulsive, and he did not resist. Her arms held him in a tight, desperate grip, and she rested her head on his chest. He put a hand on her fine, pale hair. She smelled of scented soap.

'Save me from all this. These men.'

She looked up into Vasin's face, her breath on his lips. The warmth of her slender body against his triggered an unstoppable tide of arousal. She stretched up and kissed him. Vasin felt momentarily weightless, the world spinning under him. After a moment he pushed her away, more roughly than he had meant to. He backed toward the door, wiping his mouth on his sleeve.

'I disgust you.' Masha hugged herself, shrinking.

'No, you don't. But I can't.'

He fumbled with the door handle and almost fell into the corridor in his haste to escape. Staggering like a drunk, he crossed the dining room and stood in front of the window, seeing nothing. Behind him he heard Masha slowly crossing the room. Instinctively he turned his face away, as if by his not seeing her she would disappear.

She paused by the window, pulled aside the net curtain, and glanced out at the street.

'Eyes and ears all around,' she whispered.

He took a step forward and followed her line of sight. Two watchers loitered on the opposite side of the boulevard, not bothering to conceal themselves. Vasin swore, quietly.

'Come back to the bathroom. I have something to say to you.'

Vasin shook his head. Masha moved closer to him, her voice low and sibilant in his ear.

'I know a safe place we can meet. Not even your friends out there know all the secrets of Arzamas.'

Vasin tried to form the word *no*, but his mouth, still trembling with the taste of her kiss, would not say it.

'Where?' he whispered.

'The Univermag.'

The central department store – the most public place in the city. For a moment Vasin thought that she was making a desperate joke.

'There's a cafeteria in the basement for the shop workers. A Georgian called Guri runs it. Tell him you're a friend of Seraphim's. Say I sent you. He can call me at any time.'

'Masha, we cannot.'

'You don't want to see me again?'

He ran a hand through his thinning hair. Yes, I do.

'No. I don't. Not this way.'

Masha nodded, slowly.

'Adamov used to say, "There are no secrets about the world of nature. There are secrets about the thoughts and intentions of men." Typical of him. The physical world as something knowable, the human one a mystery. But I never agreed. There is nothing secret about men. It's all about lust, ambition, fear. Mostly, fear. Like yours, now. You don't want to lose what you have. Your place in the world. Your *kontora*.'

'It's not that. Not fear.'

'I see.'

'Not fear for me, I mean.'

'For *me*, then?'

Vasin could find no answer.

'Can it be that you are really an honest man, Sasha?'

Her low tone was half-mocking. But only half.

Vasin thought of Katya Orlova. Of the dossiers buried in Orlov's safe. Of the soulless, numberless universe of brutalized functionaries that he served, a nebula so vast that it had created its own moral gravity.

'*Honest?* I don't know what that means.'

They parted in the hallway, Masha thanking him formally and

loudly for his help and his strong back for the benefit of the micro-phones, with expressions of mutual esteem. The stairs, as Vasin descended them, seemed made of jelly, the street a sea of pearly autumn light.

I am a thief of the spirit, he thought. Faithless. I seduce people into confidences and betray them. What does it mean, then, if I myself have been seduced?

Vasin felt the disconnection that usually came with drunken-ness, though without drink's sweet numbness. He remembered a childhood summer holiday at Gagra, on the Black Sea coast, with his mother. Striking out from a beach larded with pale bodies, Vasin had swum far into the swell. Beyond the reach of his mother's admonishing yells, far out at sea and as far from his fellow human beings as he had ever been or ever would be in his life, Vasin's schoolboy backstroke failed him. Cramping in a cold current, his body refused to obey. He remembered the calmness and clarity of that moment. The intense blue of the sky, the refractions of sun-light in water as he bobbed below its surface. And far away, the horizon of the world to which he was suddenly sure he would never return. Above all, his sense of childish indignation. He was in the wrong element. Looking downward, he saw the dark blue place where he would presently drown. It was inexplicable. It was unfair. He should not be here. He was in the wrong place. And death was coming to punish him for his trespass.

Then the strong arm of his rescuer enfolded him, hauling him back into air.

Vasin felt a similar helplessness now. With Masha he had plunged into the wrong element. His faculties of reason and calculation had deserted him. He felt himself transfixed in a private maze of signs and portents, unreadable to him. Most disturbing of all, this maze had appeared inside himself.

Vasin walked to Kuznetsov's apartment, trailing silence. He tried to wrestle his thoughts back to the case. This puzzle, as firm in his hands as a knot, had become the only substantial thing in his suddenly inexplicable life.

III

By daylight the corral of barracks where Korin lived looked even grimmer than at night. The gravel road had largely dissolved into the underlying mud, and the frozen ruts tripped Vasin as he walked. The knocked-together houses seemed to have given up on straight angles and stooped and tottered like drunks. He glanced at a wooden pole to which an ancient electrical junction box was haphazardly attached. Cables sagged toward the houses. The only light shone from the windows of Korin's block.

There was no answer when Vasin knocked gingerly on the door. Peering in through the window, he saw a figure curled on the bed, still fully dressed. Vasin fished in his pocket for a kopeck and tapped it on the window.

'Come!'

Inside, the barrack was nearly as chilly as the outside air. His breath steamed. Korin remained recumbent.

'Major Vasin, sir.'

'Hell you want?' Korin propped himself up on an elbow. His face was gray with exhaustion. 'Masha? Again?' Korin's voice was anxious as a father's.

'No, sir. Maria Vladimirovna is fine. But I wanted to talk to you. Urgently.'

Wearily, the old man swung his legs off his couch and rubbed his face. He glanced at the cold stove.

Korin turned his heavy, bearded head toward Vasin. 'Coffee is in the kitchen.' Korin addressed him in the familiar form, but his tone was more paternal than patronizing.

As Vasin heated the pot, Korin pushed past him into the primitive bathroom and urinated unceremoniously, not bothering to close the door. On his way back through the kitchen the old man snatched up a paper bag, grabbed a handful of biscuits, and stuffed them into his mouth, unselfconscious as a child.

He shuffled across the room and sank into the decrepit sofa in front of the unlit stove. Vasin brought two steaming mugs and sank into an equally knackered armchair opposite him.

'This better be good, son.'

'Wanted to talk to you about Adamov.'

Korin grunted dismissively.

'My job, sir. A man is dead, Colonel Korin.'

'A man is dead! A man is dead!' Korin squeaked the words in a mocking falsetto. 'What's one man, Vasin? What's one academician's son compared to what we are achieving here?'

Vasin couldn't think of any answer that didn't sound unbearably naïve.

'I have come across some information. I understand there may be certain dangers associated with the latest design. I heard that Adamov is concerned about the power of RDS-220. He's worried about the "unpredictable effects".'

Korin sat forward abruptly, like a machine that had been switched on. His eyes were intense.

'Who do you have on the hook, Chekist?'

'Apparently there was a new type of . . . "tamper"? Made of solid uranium? That could have the effect of—'

'Stop right there, boy. Your little canary has been singing you a fine song, but you have no idea what you are talking about.'

'I believe it is relevant to the death of Fyodor Petrov, sir.'

Korin stared from under his bushy eyebrows with a ferocity that could have stopped a truck.

'Have you mentioned these fairy stories to anyone else?'

'No, sir. I have not.'

'That true?'

'Yes, sir.' Vasin felt himself involuntarily quaking like a scolded schoolboy.

'Why not?'

'Because . . . it is unconfirmed. Just a story. As you say. But I wanted to ask you . . .'

'Vasin – that your name? Vasin. Listen to me. Forget everything you think you know about RDS-220. There's nothing for you there. You are crashing about in something that's more dangerous than you know.'

'That sounds like a warning.'

'It is a warning. It *is* a fucking warning. And before you say, "What have you got to hide?" I tell you this. That half-baked information you are blabbing about? It touches on the most closely guarded secrets of our Motherland. I promise you, whoever it is you work for, that your boss's *boss* isn't authorized to know this information. You understand? Just forget you ever heard it. Take it to your grave.'

'Does Adamov fear that the bomb could be *too powerful*?'

'Are you fucking *deaf*, boy?'

'Help me understand. You don't have to tell me any more secrets. But why am I being told that this is connected to the Petrov case?'

'Someone told you that?' Korin's voice suddenly cracked with urgency. 'Who? Who the hell told you that?'

'Just tell me why they're wrong.'

Korin subsided into his sofa, chewing his lip. 'Not your own theory, then? Someone in the Institute going about saying this shit?'

'Correct.'

'They told anyone else but you?'

'No, sir. Not as far as I know.'

'Just the two of you. You and your little stool pigeon.'

Vasin shrugged an affirmative. Korin breathed a silent oath.

'Very well. Let me ask you some questions, to help you understand. What is a nuclear bomb?'

'A weapon of war.'

'Just a very big bomb, a weapon of war, you say. Wrong. They were once. But now there are too many of them. They are too powerful. Nuclear weapons are the end of human history in all places, for all time.'

Vasin was silent. Outside the windows the afternoon light faded.

'And you know who wields that power? Men like General Pavlov. Come across him?'

Vasin nodded.

'Ever heard of Totskoye? Don't bother answering. You haven't. A godforsaken place out in the Orenburg steppes. Back in September 'fifty-four they tested a bomb there. RDS-4. Except it wasn't the bomb they were testing. It was the effect on men. Whose men? Our men. Yes. Forty-five thousand Soviet troops marched out into battle positions on the steppe. The 270th Rifle Division, plus about twelve hundred tanks and armored personnel carriers. They were told that it would be a regular military exercise, but with a mock nuclear explosion that would be filmed. No protective gear was given. They dropped the bomb about thirteen kilometers away. Close enough. A forty-kiloton yield, detonating at an altitude of 350 meters. About twice as powerful as the Hiroshima device. About five minutes after the blast they sent in aircraft. Three hundred of them, to drop conventional bombs on the area. Then, three hours later, they sent in the tanks, to practice taking a hostile area after a nuclear attack. And in the meantime the general who planned the whole exercise was sitting in an underground nuclear bunker. Know who it was? Marshal Georgy Zhukov, that's who. The most decorated soldier in Soviet history. The Victor of Berlin. The Victor of Kursk. He sat in a bunker while his men advanced into the hot zone.'

'Why did they do that?'

'Political games, boy. They wanted to test the effects of nuclear fallout on men. And to make a point. Play some politics. Stalin always believed that nuclear weapons could never be used on a battlefield, only deep behind enemy lines. His deputy Malenkov agreed. And after Stalin's death, Malenkov was in line to take over. He hated Zhukov, feared his reputation and his ambitions. So the honored Marshal and war hero decided to stage a nuclear test, with human subjects, at Totskoye, to prove something about a battlefield nuclear war. But mostly to prove that Stalin had been wrong, and that Malenkov was still wrong.'

'What happened to the men?'

'The *men*? Fucked, to a greater or lesser extent. Cancers. Leukemia. Radiation poisoning. You saw what happened to your precious Petrov. It was worse for the tank men who drove right to ground zero, and inhaled the heavy-metal fallout.'

'They never knew?'

'The nuclear detonation would have been hard to fucking miss. I've seen a few myself. They'd been told it was safe, and who are we to question the wisdom of the Party and government? The Party is the intellect, honor, and conscience of our country. Oh, I can tell what you're thinking. You're thinking: Korin is a subversive element. This traitorous jailbird. How can he be trusted with a responsible position in the defense of our Motherland?'

'No, sir.'

'No, sir. Not what I'm thinking at all, sir.' Again the mocking falsetto. 'Keep talking, sir, and I'll write it all in my little report that I'll submit to my superiors in due course and they'll praise me for my diligence and move me up the queue for a new car.'

Korin stilled Vasin's attempt to answer with a raised hand.

'*No.* Just listen, once more, and I will spell it out for you in simple words. Ever heard of Richard Jordan Gatling and Cyrus McCormick?'

'Never, sir.'

'They're Americans, boy. Engineers. These days they tell the kids that Russians invented everything under the sun. Ridiculous. McCormick invented a reaping machine, boy. Back in the 1830s, when Pushkin was alive and we were serfs. It could do the work of a thousand men with sickles, and never tire. And Gatling invented a gun. Based it on a seed drill he'd designed. He discovered that feeding seeds into a pipe or bullets into a chamber is the same principle. Gatling designed the first machine gun. Ten barrels, hopper-fed, hand-cranked. It fired two hundred rounds a minute, more than an entire battalion of soldiers with muzzle-loading muskets. Mowed men down like standing corn when the Americans had their civil war. McCormick invented a reaper of corn, Gatling a reaper of lives. I read Gatling's memoirs back when I was at the Kharkov Institute. He said his machine gun bore the same relation to a rifle as Mr McCormick's reaper did to a scythe or Herr Singer's sewing machine to a plain needle. And he was right. Automation could multiply the work a human hand can do a thousandfold. Gatling's genius was to bring automation to war. And he did it because if men had such terrible weapons, they would see the futility of fighting war. If a four-man machine-gun crew could kill a thousand infantrymen in five minutes, what would be the point of ever fighting? That's what our Gatling thought. And was he right?'

Vasin shook his head obediently.

'Automated war just meant bigger slaughter. And Robert Oppenheimer. Another American. He built the first ever atom bomb. Called it Trinity, like the heathen Jew he was. He said, "The atomic bomb has made the prospect of future war unendurable." Sound familiar? He said, "It has led us up those last few steps to the mountain pass; and beyond there is a different country." That was just after they dropped two devices on Japan. Oppenheimer thought using atom bombs on cities would make war unthinkable.'

'Last time we spoke, you said RDS-220 is about stopping war forever.'

'My friend remembers. The difference is that Oppenheimer never reached his "different country". The largest bomb he made was only twenty kilotons. That's as much conventional explosives as a thousand heavy bombers can carry these days, more or less. He multiplied the killing work a single bomber could do by a thousand. See what I'm getting at? Oppenheimer was just another Gatling. He made war more deadly, but didn't make it *unthinkable*. Human minds – or at least the minds of generals – don't work like that. I told you about Totskoye. As soon as we had our own atom bombs, we immediately began thinking about how to use them on the battlefield. No. For war to be truly *unthinkable*, it must be truly *unwinnable*. We need a weapon so powerful its use is suicidal for everyone. "A thermonuclear war cannot be considered a continuation of politics by other means. It would be a means of universal suicide." At least you know the man who said that. It was Adamov. And it's Adamov who has brought us to that new land of Oppenheimer's at last. We've reached the final thousand-times multiplication. Adamov has discovered the means of mankind's extinction as a species. That postwar place that Gatling and Oppenheimer dreamt about? RDS-220 will put us there. Science is on the verge of vanquishing war. The intellectuals will have beaten the soldiers at their own game. Finally, we will have a weapon so deadly that even the most pigheaded general would never use it.'

'Only if RDS-220 is a success.'

'Fucking right.'

'But why would Adamov alter the design of the tamper?'

Korin cut him off, suddenly furious.

'Who are you to ask *why*? Tampers and uranium. You are an ape poking about in a laboratory. You are as lost as the soldiers and politicians in charge of this whole place. None of you have any idea of what you are dealing with. Science will forge a different country. *Adamov* will, do you hear?'

Korin leaned forward and grabbed Vasin's upper arm with a grip that could have burst a bottle. The older man shook Vasin's whole body, twice, hard. Then he released his grip. For a while the only sound was Korin's labored breathing.

'You are saying, Colonel, that this is bigger than me. That I am not authorized to continue doing my investigation, nor am I authorized to know why.'

'If you haven't understood the stakes by now, friend, you never will.'

'Let's say I do *not* fuck off out of your affairs. What will happen, Colonel?'

Vasin was speaking in a low, controlled voice, but he knew that it was his stung pride talking. He had expected his insolence to spark another flash of anger from the old man. But instead Korin only swung his legs back onto the couch and sank deeper into his stained pillows, as though the fight had gone out of him. Or perhaps he had lost interest in arguing.

'Then there is nothing I can do for you.'

Korin passed a hand over his weathered face and closed his eyes.

The old man's tongue flicked over his lips, like a lizard's. After a long moment he opened his eyes again. Korin's former passion, his storyteller's animation, had disappeared. His stare had become hard and cold.

'*The tongue hath the power of life and death.*'

'What did you say?'

Korin had spoken the phrase in Church Slavonic, the ancient language of the Russian Bible. He turned his face away from Vasin. His voice came as a muffled growl from within the pillows.

'You heard me, boy. *The tongue hath the power of life and death.* And you have spoken.'

PART THREE

SCOURED, MELTED, AND BLOWN AWAY

The ground surface of the island has been leveled, swept and licked so that it looks like a skating rink. The same goes for rocks. The snow has melted and their sides and edges are shiny. There is not a trace of unevenness in the ground . . . Everything in this area has been swept clean, scoured, melted, and blown away.

REPORT OF PHOTOGRAPHIC TEAM SENT
TO EXAMINE EFFECTS OF THERMONUCLEAR
BLAST ABOVE TEST FIELD D-2, MITYUSHIKHA BAY,
NOVAYA ZEMLYA ISLAND, 1961

CHAPTER EIGHT

FRIDAY, 27 OCTOBER 1961

THREE DAYS BEFORE THE TEST

I

So many secrets, buried in Russia's earth. Vasin had been down to see them. An underground city of files, arranged in stacks on steel shelves, stretching into darkness. An acid smell of old paper and a faint, moldy tang of decay. The KGB's Central Registry, in the Lubyanka cellars, was the graveyard of a million sins and betrayals. Except unlike in a graveyard, Vasin could tug these small tombstones off the shelf and, Godlike, learn the dead's forgotten stories. He could read of their struggles as they fought to stay alive, to walk under the sun, raging against their burial in deep, secret places.

It had been nearly two years ago, Vasin's first month at the *kontora*. Armed with a fistful of authorization slips, duly stamped and signed, he had descended in a service lift to the bowels of the Lubyanka. Orlov could have ordered the files up to his office. Instead he had sent Vasin down to find them himself. If it was meant to be a lesson, its meaning was still obscure to Vasin. Behold how much treachery there is in the world to be rooted out? Take a walk in our mausoleum of memory, know that for every corridor there are a thousand more? See how many corpses we have made?

Vasin had stepped out at the third basement level. The passageway was low, its ceiling cluttered with wiring and pipes that twisted against each other like veins. When he swung open the door to the registry, the desk clerk's eyes flickered in alarm, as though visitors from the upper regions carried some dangerous virus. The woman had the pallor of a cadaver. She took his slips without a word and copied the file numbers, slowly, into a register.

'First time?' she asked in a gravelly voice. Vasin nodded. She blew out air, as if she were smoking, and stood. 'Follow me.'

She led him through tunnels of paper, endless as a labyrinth. Every wall was stacked with rows of files, all neatly labeled, interspersed with gray steel cupboards and shelves full of old red-bound books. There was a scent of sweet dust and black tea.

'Nineteen fifty. Category 151 cases. It will be here. Leave the slip on the shelf in place of the file, recover it when you replace the dossier. Reading room is down there.'

Vasin found the documents without difficulty. The boxes' spines had been stamped with consecutive serial numbers in heavy black ink. When he slipped the correct box from its place, the file was so heavy he had to cradle it like a baby. It sat in his arms, eerily malignant, a swollen tumor of paper. Three kilos of paper that equaled a human life.

The reading room was a long, low hall lit with hanging steel lamps and furnished with standard-issue office desks. A dozen men and women, all in uniform and all pale as stationery, pored over boxes like his own with a concentration so profound that they seemed bewitched. Nobody looked up as he entered. The silence was broken only by the rustling of paper and the scratch of pencils.

It had been Vasin's first job for Special Cases. Orlov had started him off on something familiar: bribery and extortion. Nikita Olegovich Belov, secretary of the Regional Committee of Ivanovo Province. A midlevel Party apparatchik, no vices beyond the usual, on paper a typical example of the species. His official photo showed a bloated face with a bully's straight gaze and the resentful pout of a much-commanded man. Like most officials of his level, Belov's rise had been punctuated by a flurry of official complaints from colleagues and citizens, swirling in his wake like candy wrappers in the backwash of a Party sedan. Complaints, the tiny revenge of the powerless. Everyone wrote them. Tales of drunkenness and domestic cruelty, accounts of the indifference of post office clerks and the slovenliness of meat-counter workers, accusations of petty theft, adultery, and espionage. Encouraged by officialdom as a harmless way for citizens to vent their frustrations, these complaints were the bane of newspaper editors, foremen, and policemen. And useful, too, on occasion. Troublemakers carelessly flagged themselves to the authorities with their scribbled indignations. And denunciations formed a helpful store of petty transgressions that could be produced to cut short the career of an enemy or anyone who had become politically inconvenient.

'Every man goes through life carrying a sack of sins, Vasin,' Orlov had told him. 'As Comrade Stalin said, show me the man and I will tell you his number in the Criminal Code.'

And so it came to pass that Special Cases had decided, for reasons Vasin never presumed to know, to send him to examine Comrade Belov's personal sack of sins. A particularly unattractive bagful, as it turned out. His progress up the greasy pole had been, by the accounts of the hysterical widows and passed-over colleagues, facilitated by blackmail. A serial denouncer, Belov, and one who, according to his accusers, had demanded bribes in return for his silence. Those who refused to pay had been reported. Back in the days when a Party official's word was more than enough to condemn a man to the Gulag, Belov's name had cropped up on an unusual share of damning reports and deadly memoranda to the *kontora*. It hadn't mattered until suddenly, in Orlov's judgment, it did. It was Vasin's job to comb the files and gather the facts.

The first box on the list of Belov's victims was the dossier of one Slutsky, Foma Petrovich. Party worker from Novgorod, wife and three children. Condemned to die on 15 October 1950, for sabotage and spying. At the time, Slutsky had been Belov's equal at the local Party branch, and therefore a rival.

When Vasin pulled the black cloth tag that held the box closed and opened its crumbling cardboard cover, the papers inside exuded a slightly acidic musk. Most of the file's pages were flimsy official onionskin forms, punched through in places by heavy typewriting. Interspersed were a few slips of thicker, raggy scrap paper. Toward the end were several sheets of plain writing paper covered in a thin, blotted handwriting, Slutsky's confessions to being an enemy of the people. Vasin, exhausted by the enormity of his task even before he had begun, settled down to read.

The file existed on the border between banal bureaucracy and painful poignancy. It was a compilation of the absurdly petty – a receipt for the confiscation of Slutsky's Party card, the confiscation of his daughter's Young Pioneer holiday trip voucher – and the starkly shocking. Long confessions, in microscopic, crabbed handwriting, covered with blotches and evidently written under extreme stress. Thick wads of testimony, cross-referenced and typed verbatim, given by Slutsky's accusers and by his fellow defendants. Belov's damning indictment of his colleague as a traitor and a spy. Transcripts of the court hearing. The conclusion of the three-judge court, first in long-hand and then typed. The verdict of death seen and signed for, with bureaucratic neatness, by Slutsky himself. And finally, at the end of the file, a slip with the stamp of the Novgorod prison and a scribbled signature verifying that the sentence had been carried out.

At the end of that first day, the passing of time marked by no change in the hard electric light but only by the soft whisper of files being closed and chairs shifted backward across the linoleum, Vasin closed the bound stack of papers with an exhausted sigh. Before he shut the box's lid, he stared at the fussy cursive of the first docu-ment, the seed from which the rest had sprouted like a poisonous plant. Belov's handwritten denunciation, on a small piece of expen-sive blue notepaper. 'I beg to approach you, Comrade guardians of State Security, with some disturbing information that as a loyal Soviet citizen I am bound to report . . .'

In all, Vasin had spent three weeks in the Lubyanka's catacombs. Down in these tombs of secrets it began to seem to him that no weather, no sunlight, no open space existed or had ever existed. Emerging into the chilly air of the spring nights, he felt as though he had woken from a deep fever-sleep. Stepping back into the lift in the mornings was like clambering back into his own grave. Every box file he touched and every cup of bitter tea he sipped seemed charged with suffering, like electricity. Sitting in the quiet of the reading room, he began to imagine, like a man hallucinating, that the whispering of the water in the central heating pipes overhead was the murmur of dead human voices. Though in truth the sound of these files, if they could make sounds, would be a scream.

He came to know Belov's crabbed handwriting well. A lurch, like that of the Ferris wheel in Gorky Park that had so terrified Nikita, would grip him every time he saw the fatal denunciation. This man had written the death sentences of dozens. And, just as Orlov had known, the evidence of his venality was all there for those who chose to look. A desperate victim's wife, panicking in the interrogation room, protesting that Comrade Belov had demanded a five-hundred-ruble bribe to refrain from his accusation. A daughter, freshly in shock after her expulsion from university following the arrest of her father as an enemy of the people, sobbing a barely coherent story about rejecting Belov's sexual advances and ignoring his threats to destroy her family.

Inevitably, as he turned the pages of each dossier, Vasin would learn from the terse typed-up referral slips that most of these accusers had themselves been punished in their turn. 'Slander of a public official.' 'Antisocial tendencies.' And, more than once, 'mental instability.' A suspicion of slow-onset schizophrenia, in the investigator's opinion, 'subject to be referred for psychiatric evaluation.' Somewhere, in some other cellar in some other building, Vasin knew the rest of the victim's story lay carefully recorded by the all-seeing eye of Soviet bureaucracy. Assessments, medication, commitment to psychiatric care. Vasin knew how that ended, an agony of terror on twisted, shit-stained sheets, like his sister, Klara. Or else the story would continue in the files of a labor camp or a prison or an exile settlement somewhere deep in the Kazakh steppes. Illness, suicide, lonely death. A universe of suffering, recorded in a million words dropped one by one into silence.

Vasin remembered how he had climbed into bed beside Vera during those weeks. He had touched her shoulder but she didn't stir, so he knew she was awake. But he could think of no words for what was going through his mind. When he finally found sleep he was powerless to stop his mind from filling the day's words on the page with flesh and blood, from giving them voices and tears.

Sentimentality? Not something he had ever suspected in himself, not after years as a Moscow criminal detective. He had spent his career in society's underbelly, as elbow deep in the shit and the gore as any morgue orderly or sewage man or abattoir worker. He knew the sour stink of prisons, so pervasive that you tasted it in your mouth for days. He knew about the mutilations and the sodomy and the sordid ways men chose to kill and die in the hopelessness of their captivity. A straitjacketed prisoner in Moscow's Butyrka prison who had bitten his tongue in the hope of choking on his own blood. A suspect who had tried to drown himself in the latrines, kicking at the guards as he desperately fought to inhale sewage. He had seen the netherworld where the greatest horror of all was to remain alive. More, he had condemned perhaps hundreds of men and women to life, and sometimes death, in this underground.

Men had made a hell for the guilty. But what did it mean to send the innocent there, deliberately – or even worse, through simple carelessness? Something in Vasin rebelled. His childhood sense of fairness was offended. It seemed too feeble an instinct to stand against the storming fury of so many unquiet ghosts. But Vasin knew he had discovered anger. He could smell its first kindling and hear the crackle of its brushwood. We must cut down the serf inside ourselves, he remembered some great Russian writer saying. We must set our good hearts free, which is the dream of every decent soul. And even – it must be – of certain serf masters, too.

When Vasin had stood before Orlov, at attention with report in hand, the General had taken in his pallor, his pinched convalescent's face. What was the lesson that Orlov had wanted him to learn? Did his boss even believe there was a lesson to be learned, down in that world of the dead? Perhaps, thought Vasin, a man who had spent a lifetime inside the system had no knowledge of its enormity, like some blind cave fish who knew nothing of the sun? But Orlov had said nothing. Or nearly nothing.

'Did you find what we were looking for?'

'Yes, sir. Comrade Belov is corrupt, without a doubt.'

Orlov silenced him with a hand, impatiently extended for Vasin's report. His boss flicked through it, grunting occasionally.

'Good work.'

'Thank you, sir.'

Orlov caught something in Vasin's tone. His head snapped around, like a listening bird hearing a rustle in the branches.

'Vasin.'

'Yes, sir?'

'*Vasin?*' This time a rising note of question.

The General paused, perhaps weighing if he had the time to make a speech, and decided against it.

'When you fell trees, the chips fly.'

'Yes. Sir.'

Vasin struggled to choose words, but Orlov had already turned away and was rooting in the pocket of his breeches for his keys. The green steel safe, the first time Vasin had ever seen his superior open it.

'That will be all, Major.'

The next month, Vasin read in *Pravda* that Nikita Belov had been accepted as a candidate member of the Central Committee. In the *kontora* canteen they spoke of him as the next Minister for State Security.

So much poison, dug into the ground. So much betrayal and viciousness trapped in those lifeless pages. All those dead, their cries for justice composted into poison that could seep out like marsh gas. Even a tiny waft of it, released into the outside air, could kill. Academician Petrov's secret, for instance. Had Adamov known all along, as he sat in the courtroom and watched his wife denounce him as a traitor, that his friend and colleague had brought him to this pass? Could Adamov have found out more recently? Had Fedya Petrov, so bright and clever and vital, been murdered by a shade of the past escaped from its deep cellar and now abroad in the world?

Through the thin partition of the Arzamas apartment, Vasin

heard Kuznetsov loudly clearing his throat in the bathroom. Then the shower pipes, shuddering into life with an equally throaty gurgle. Outside the drawn curtains and the closed door, the new day stood ready for him.

Arzamas buried its secrets, too. Somewhere deep underground and very close, men like Adamov were moving among their stockpiles of deadly metals, building their infernal machine, patiently constructing Armageddon. But the *kontora*'s own poison cellars were stocked with a knowledge no less deadly, waiting to burst explosively out into the light.

II

Vasin sat on a hard wooden bench in front of Adamov's office, waiting for the Professor. Outside the Institute an enlarged detail of Zaitsev's watchers waited in turn for Vasin. But some unwritten protocol kept them outside the hallowed halls of the Institute itself. This building had become the last place in Arzamas where Vasin could move around without close surveillance.

On the wall outside the Professor's office was an arrangement of official photographs of the luminaries of the Institute. They all wore the hard glares of men who stared through hardship and into a glorious future. The faces in the upper row were underlined with a constellation of stars, spreading across their tunics. Hero of Socialist Labor. Hero of the Soviet Union. The portraits of the deceased bore black ribbons across one corner. In the fourth row Vasin found Fyodor Petrov, his ribbon not yet added. Fedya, still alive on the wall, had worn thick-rimmed glasses for the shot. Another vanity, guessed Vasin, albeit one of false modesty. He had wanted to render his matinee idol's face more serious, for the eternal record. His white lab coat bore no medals. And now never would.

Vasin sought out Adamov, in the center of the top row. Stern, like the rest, formidable. Resolute. The photographer had found no trace of humanity or humor in the sharp lines of that countenance. Vasin could stare at the photograph in a way he would never dare at Adamov's live face.

Did Adamov dream of the Gulag? Vasin wondered. Did the faces of all the men who had ever caused him pain come to him at night, the swine-faces of the camp guards, the pale ratlike faces of the hypocrites' court that had condemned him? Petrov, the once-beloved colleague who had betrayed him so ruthlessly? His faithless first wife? His daughter, who never wrote, even after his release and rehabilitation? The two women must have renounced Adamov in their righteous indignation at his supposed treachery. Or was it fear, and now they remained far from him out of shame? What lies did they weave in their minds when they thought of him? Could Adamov justify what they had done, even forgive them? What lies did he tell himself about their reasons? Did the man keep a spark of love alive for them, nestled deep in the ashes of the past in the cold grate of his heart?

Vasin tried to imagine Adamov, this brilliant, imperious man, in the hard places of the world. A golden mind, the Russian phrase went. And at the same time, he was also a human skin, a bag of hormones and organs, a sack of desire and digestion. Vasin saw him freezing in a thin prison uniform, starving and pushing forward, crazed for food, breathing the stench of other prisoners, his hungry belly filled only with the stink of their common humanity. Adamov slurping thin soup from a flimsy metal bowl. Fighting for a place to squat in the fetid dark of a prison latrine.

Vasin thought of Butyrika prison, the faces he had seen peering through the hatches in the battered old doors. Most were uncomplaining men, doomed to be abused, men to whom things simply happened. Their grandfathers had been someone's property, serfs, and so on back for a hundred generations. These men had inherited the bovine resignation of ancient Russia. But there were others, men whose reason rebelled. You could tell them immediately. A snap in the glance. The way they straightened when the handcuffs came off in the interrogation room. A lingering dignity in the spread of their shoulders. The prison guards saw it, too, of course, with their scavengers' instinct for scents of weakness, or dangerous strength. Proud men always had it hard, at the beginning. Then, in time, their fellow inmates would accord them a grudging space, as a race apart.

How had he become a man who knew such things? wondered Vasin with a sudden flash of revulsion. This knowledge of the degraded subtleties of prison society. How had he become a gate-keeper to this filthy world? Its interpreter?

'Mama, I am going to be a policeman.' Even his mother, this woman for whom the Soviet State could do no wrong, had flinched at the idea. The way she had paused, halfway across the communal kitchen, taken a moment to compose a brave smile.

'It's not like Uncle Styopa, you know.'

Uncle Styopa, the kind neighborhood cop in the patriotic children's poems that she had read to him. Tall, blond, and strong, Uncle Styopa had prevented train crashes, helped firemen, punished a school bully. He was a fighter against injustice and a hero to all Young Pioneers. She knew him as only a mother could. Of course Uncle Styopa was exactly what Vasin wanted to be. And even, then, what he thought it would be like.

'It will be an honor for you,' his mother had said, eventually. 'Your father would be proud.'

Vasin had no way of knowing how his father would feel. He remembered a warm, ironic smile, the feel of the rough cloth of his father's uniform tunic on his childish face. A week or so before his father had left for the front for the last time, his mother had taken Vasin to watch the troops parade. She had waved excitedly as his father's unit passed, and the boy Vasin had waved and cheered too. But the truth was that he had not been able to recognize his dad among the rows of rhythmically high-stepping soldiers. It was as though the man had already been absorbed into the khaki mass, dissolved into the collective. His father had become part of a machine, an unrecognizable dot in a sea of identical faces.

In death, Vasin's father had been doubly taken from him, no

longer a private man but transformed into an archetypal war hero. One of the glorious dead who were everybody's common property. His father's apotheosis had automatically made him a patriot, a mute statue onto which the world would hang its placards.

'He defended our Motherland,' his mother concluded, finding a trite label for her feelings as usual. 'Now it is your turn.'

Vasin had known he made a mistake the moment he walked into the Police Training Academy. Not an escape from school but a nightmarish replay of it. The shaven-headed, low-browed faces of his classmates, scowling at him with an instinctive knowledge that the new boy was not one of them. A smart one, this. Thinks a lot. Even his instructors had mocked him. Cadet Vasin, you know all the answers. Vasin will know this, he has spent the vacation reading the *Great Soviet Encyclopedia*. The plodding sarcasm of a mediocre world. Only the women instructors who taught forensics and fingerprinting had admired him. Always so smartly turned out. An example to you all, boys. The others had sniggered and picked their noses.

Vasin struggled to wrench his thoughts back to the present. He concentrated on the dust dancing in a beam of morning sunlight that pierced the blinds. Particles rising and falling in space, in eternal random motion. God's feeble way of entertaining bored physics students.

Adamov and his entourage appeared at the end of the corridor, the great man himself striding in front and his coterie of white-coated assistants following behind, moving with great urgency. Vasin stood. Adamov's plain brown suit was stained with chalk, and his knotted tie was askew. His face wore its usual mask of stony seriousness, but the eyes were exhausted. The Professor fixed Vasin with a beady stare. As Vasin stepped forward into the Professor's path, the white-coated assistants moved to surround him like white blood cells attacking a germ. A burly lieutenant colonel in an engineer's uniform body-blocked Vasin.

'Not now.' The officer's voice was hoarse and his eyes rimmed with red from lack of sleep. 'Get lost.'

'Professor,' Vasin called over the man's shoulder. 'A word in private. It's important.'

Adamov stopped and blinked in the neon glare of the Institute corridor, a bright white against the dull early morning's effort beyond the windows. The nervous energy of the group focused into a jittery hostility at Vasin's insolence. An older scientist in a stained lab coat spoke up.

'The Professor has no time. We have important work to do.'

'Two minutes, Comrade Professor. What I have to say is most urgent and concerns you, personally.'

Adamov did not accord Vasin the courtesy of going into his office. He merely swung open the nearest door and with a scowl sent the occupants scurrying for the exit like cockroaches in the light. As the room emptied, Adamov scanned some documents on a clipboard he held, as if bothering to look Vasin's way would be a waste of a precious second.

'Speak. The device must be ready for transportation by the end of tomorrow.'

The Professor whipped a pencil from his breast pocket and made a note on his clipboard. Some men had a gift of quiet. In their presence, intimacy grew, and tendrils of doubt and confession like softly climbing plants. Adamov was the opposite. His silence was violent, a storm of tension building like a charge in a machine that would soon emit a devastating snapping arc of static electricity.

'I know Academician Arkady Petrov denounced you in 1937, Professor.'

Gradually Adamov's stillness became awesome. He turned his head away, and the light fell on the square cage of his temples and jaws and the trapped and furious eyes within it. The terrible power of secrets. It had given Vasin power over Masha and now, over this great man. He felt shame. It was like knifing a man in an honest fistfight. Dirty dealing. He was not equal to the weight of the weapon he held in his hands.

The Professor's answer, when it eventually came, seemed to echo from far away.

'Arkady Petrov.' Nothing in Adamov's flat repetition gave any clue as to whether this was a confirmation of what he knew or a revelation. 'With whom have you shared this information?'

'Maria Vladimirovna. Would you like to know what she said?'

Masha's shadow fell between them like a sword. Vasin remembered the flutter of her eyelashes, the knowledge thudding into her chest like a blow from a club, the weight of her body against his.

Adamov exhaled. He looked at Vasin as though truly seeing him for the first time. In the crowded world of Adamov's mind, Vasin imagined that he had finally broken through the screen of the insignificant to become real in the Professor's eyes.

'*Why?*'

There was genuine pain in Adamov's voice. So Masha was the key to him. Far adrift in his ocean of abstractions, Adamov still had love in him. He had not yet reached the freezing point of absolute loneliness. Only the man with love in him carries always in his heart this capacity for damnation. Vasin saw it, and hated himself for seeing it.

'To ask her if you could kill Fyodor Petrov, the son of the man who denounced you. If you were capable of poisoning the man during dinner at your own house.'

Adamov deflated, a man worn thin as paper.

'And what did my wife say to you, Major Vasin?'

'She said that you were incapable of such violence. But . . .' Vasin forced himself to press on. 'I also know that immediately after the death of Petrov, you changed a specific part of the design of RDS-220. The tamper, designed by your dead colleague. Why?'

A quiet thickened between them. But this was not a silence of hostility. Now their time, the measure of their silences, belonged for a moment to Vasin. The Professor took a long moment to reassemble himself, to recover his authority. Their eyes locked once more. Adamov had donned a poker face a corpse would envy. But his moment of vulnerability had made Adamov's voice softer, his adamantine carapace temporarily set aside. Though Vasin remained in no doubt that, at his core, his adversary was as hard as flint.

'You are not a fool. I'll grant you that.'

'Your answer, Professor.'

'You cannot see what we see. There are matters here which you may not comprehend.'

Indignation erupted inside Vasin. These were also Orlov's words.

' "We know better than you"?' Vasin's voice had chilled. ' "We sit higher, we see further"? Professor, you sit high and see far. But from where I sit and from what I see, you had the motive and the opportunity to kill Fyodor Petrov.'

Before Vasin could continue, Adamov, still looking away, began to speak.

'Young man. In the middle years of my life, after the camps and before I was allowed to return to Leningrad, I lived on half a hectare of land in Tambov Province. A peasant's wooden cottage with a vegetable patch fertilized with my predecessors' shit, which I dug out from the latrine. There was a meadow, completely enclosed by forest. It teemed with creatures: deer, squirrels, songbirds, crows, mice, a hawk. Except for the crows and the hawk, the carrion bird and the predator, every one of those animals constantly and fearfully watched over its shoulder lest it be caught, torn, and eaten alive. From the animals' point of view, my little paradise was a space of violence and death. Only very rarely does an animal living under natural conditions in the wild die of old age.'

Adamov had let his arm dangle free, the file held so loose in his fingers that Vasin thought he might let it drop to the floor.

'What do you mean, sir?'

The Professor continued talking as though Vasin had not spoken.

'In nature every creature lives its life in a permanent state of terror. Terror, and vigilance. To live to the end of every day he must fight to find sustenance, and to avoid death. Even the predators must kill or themselves die. The carrion birds depend on the murders of others. Why do we assume that the human world must be different?'

'I don't understand.'

The Professor suddenly seemed to become aware once more of Vasin's presence and turned his head to speak to him with a terrifying directness.

'What do we mean by "understanding" something? We can imagine that this complicated array of moving things which constitutes "the world" is something like a great chess game being played by the gods, and we are observers of the game. We do not know what the rules of the game are; all we are allowed to do is watch the playing. Of course, if we watch long enough, we may eventually catch on to a few of the rules. The rules of the game are what we mean by "fundamental physics". But even if we know every rule, what we really can explain in terms of those rules is very limited. Because almost all situations are so enormously complicated that we cannot follow the plays of the game using the rules. Much less tell what is going to happen next. We must, therefore, limit ourselves to the more basic question of the rules of the game. If we know the rules, we consider that we "understand" the world. And do you know the rules? Any rules? I doubt it. So I may say two things to you now with certainty: You do not comprehend what happened then and you do not comprehend what is happening here, now.'

'Then enlighten me.'

'You have dug up a single piece of ancient history from the files and you are spinning tales. That is all. What is your evidence for the rest of what you suggest? There is none.'

Vasin hesitated for a moment too long. The evidence that could convict or exonerate Adamov lay below them, deep in the vaults of the Institute, among the laboratory records of Petrov's experiments, which Axelrod insisted had been faked. But Vasin could not put his hand on them. And to ask Adamov for permission to view them would be to invite their immediate destruction, if the Professor was indeed guilty.

'You're wrong, Professor Adamov. Your colleagues . . .'

'Have denounced me? Again? Is this your evidence? Major, this folly must go no further. You must stop. Now. But perhaps there are some things that you need to know. So, Comrade Not-fool, I offer you this. I have no more time now. By tomorrow night we will have completed the final assembly. Come to my apartment. We will have tea and I will tell you some things that you will find illuminating. But in the meantime you will not mention these accusations of yours to anybody. Especially none of your damn bunglers in the *kontora*. To nobody at the Institute. You will keep your silence for a day. Your word?'

The image of Adamov, Maria, Korin, and Petrov sitting under the pooled light of the Adamovs' dining table on the final night of Petrov's life jumped into his mind. They had all been drinking tea.

'My word, Professor.'

III

From the corridor, the computer room sounded more like an industrial laundry than a temple of high technology. When he swung open the double doors, Vasin was assaulted by a wave of heat and noise. A long hall was filled with banks of steel-cased machines, each composed of decks of glowing electrical valves. Banks of lights flickered on the displays, and the place was hot as a sauna. Overhead, giant air-conditioning ducts poured in cool air, the pumps thrumming like a ship's diesels.

Along one side of the room were a row of glass-walled offices. Vasin spotted Axelrod gathering up a printout many meters long. He tapped on the thick glass, wasn't heard, and then banged on it with the palm of his hand. Axelrod, turning, went white at the sight of him. Like a cartoon character he turned left and right, but there was nowhere to hide in the glass cubicle. Axelrod dropped the spooling roll of paper and made a desperate gesture, flapping his hands to shoo Vasin away. Shaking his head frantically, he mouthed the word *no*.

Looking about him, Vasin spotted a group of white-coated men huddling over one of the machines. One of them was pulling out a tray from the computer with heat-proof gloves. The section trailed wires like the arteries of an extracted organ. Nobody was paying them any attention. Vasin turned back to Axelrod and dropped his hands to his fly, mimicking pissing. With a swing of his head he summoned Axelrod to follow him.

In the men's lavatory the two of them stood in awkward silence, side by side by the washbasins, as they waited for someone to finish noisily defecating. Eventually an elderly engineer lumbered out of the cubicle, smiling in comradely greeting as he took his time washing his hands. Vasin waited for the man's footsteps to recede down the corridor before turning the taps on full.

'Why did you come here?' Axelrod said, his eyes flitting around the empty lavatory.

'You told me you were running some tests. What have you found?'

'I'm being followed.'

'Calm down.' Vasin tried to keep the impatience from his tone. Everyone in Arzamas was being watched by someone. 'What's happened, Axelrod?'

'Last night, someone followed me all the way home. He was sloppy and kept ducking behind trees whenever I turned around. I don't think he was . . . one of yours. From the *kontora*.'

'It could have just been a thief.'

'A mugger in Arzamas? Are you insane?' Axelrod struggled to catch his breath. 'Who did you tell about the tamper?'

'Nobody. I have told nobody, Axelrod.' Only Korin, and Adamov.

They both froze into silence as the door swung open. The young man, seeing them locked in intense conversation, hesitated before entering.

'Get lost.' Vasin flicked his red ID card from his pocket and flashed it at the boy like a talisman.

The door slammed shut on its powerful spring.

'Vasin, I can't be seen talking to you.'

'Nonetheless, I need to know what you have discovered.'

Axelrod puffed in exasperation.

'My computer simulation will be done in forty minutes. The first set of results will prove how seriously Adamov has sabotaged the device. Come to my apartment on Builders' Street in one hour.'

When Vasin stepped out onto Kurchatov Square, a single Volga was parked by the curb, the driver reading a newspaper while the passenger kept lookout. Another car would probably be circling the square. Vasin loitered for a moment to make sure that the *kontora* men had spotted him, then strolled to the corner of Marx Boulevard. The traffic policeman, pleased to have someone to talk to, gave him overcomplicated directions. The Volga idled by the curb, then kept pace with him, trailing at a distance.

Vasin passed women in head scarves hurrying to do some lunchtime shopping. A pair of young men in new suits greeted each other with a bear hug. On the boulevards young mothers gossiped as they rocked prams. At the window of a shoe shop, Vasin stopped to check the display, watching for someone to double back or pause to tie a shoelace. Nobody did. No followers on foot, then. A steam whistle sounded from the railway, and from a courtyard came the whoop of boys playing football. Vasin walked mercifully incognito in his civilian clothes through an ordinary Friday afternoon in a city that was anything but ordinary.

Builders' Street was broad and tree-lined and crossed 8th March Street at an oblique angle. It was halfway to the airstrip, and Vasin watched a transport plane lumber into the sky over the rooftops, waiting on the corner for the Volga to nose into view. The last thing he wanted, for now, was for the *kontora* men to lose him. Axelrod's house had been built after the war from red brick. 'German buildings,' citizens called them, because so many had been put up by German prisoners of war. Even as slave laborers, the Fascists were known to build well. The facade was a wide expanse of windows pierced by a high archway that led into the courtyard. All the building's staircases and doorways opened onto the yard, and there was only one way in or out. Good. There was a children's sandpit in the center, empty of children but populated by a small crowd of oversize cartoon characters rough-hewn from wood and garishly painted. A large timber mushroom that was part of the playground furniture was usefully screened from the archway by a near life-size statue of Karandash the clown. Vasin settled on the mushroom to wait. He could observe all four entranceways from a single spot.

Vladimir Axelrod. The summary from the personnel files, pinned to the transcripts of Axelrod's interrogation by Zaitsev, had concealed as much as it revealed. Twenty-nine years old. Jewish family, father a good Bolshevik. Brilliant mathematician, studied theoretical physics under Adamov in Leningrad after the war. A stream of commendations. Worked with Fyodor Petrov here at Arzamas-16 for the last five years. Unmarried. Like Petrov, politically irreproachable, on paper at least. So why had Efremov spoken of him as politically unreliable? Probably because Axelrod had lent his friend banned French philosophical literature. Maybe that had been enough to send the likes of Zaitsev and Efremov into frenzies of suspicion. And Vasin knew that he had wept like a girl when speaking of Petrov's death. More likely than not, the tears of a lover. And equally probably Axelrod had taken the pornographic photographs and attempted blackmail. But a killer? Vasin went over the grainy photographs in his mind, the fuzzy image that might have been Axelrod. The anger in the punched-through type. *HER OR ME?*

From Gogol Boulevard came the sounds of the folding doors of an electric tram opening and closing with a snap. A minute later a tall young man came hurrying through the archway. Vasin recognized the sloping, urgent gait immediately. Axelrod disappeared into the building. Vasin checked his watch and stood stiffly. Just as he was about to step out to follow Axelrod inside, he caught a movement by the archway. He ducked back behind Karandash's shoulders. A squat, simian figure peered furtively into the courtyard, then retreated. The man wore a padded black jacket, like that of a road worker, and a cheap rabbit-fur hat. The man was trying far too hard to be inconspicuous to be a professional.

Not one of the *kontora*'s regular watchers. Someone else, with different intent. One of Efremov's irregulars? Some deniable outsider? An underworld killer drafted in for what the *kontora* liked to euphemistically call a wet operation? Vasin remembered Efremov's menacing glance up from the courtyard after he had emerged from Petrov's poison-filled apartment.

Keeping his hand to his face as though scratching his temple, Vasin sauntered from his hiding place into Axelrod's entranceway. Inside the hall he doubled back and peered from the gloom of the stairwell toward the arch. Their watcher had disappeared.

Apartment 211 was on the fourth floor. Axelrod opened the door immediately and peered down the stairwell to see that Vasin was alone.

'We don't have much time before they miss me at the laboratory. Come in.'

Axelrod's place was luxurious by any standard, but it was not the kind of sleek *nomenklatura* apartment in which Petrov had lived. It was filled with the typical clutter of the intelligentsia, floor-to-ceiling bookshelves, an upright piano, a spread of sentimental paintings of Russian landscapes across one wall. The sitting room was furnished with solid prewar furniture. Axelrod evidently hadn't made the distribution list for the trainload of Czech buffet suites that graced Kuznetsov's apartment.

Axelrod moved a pile of journals from an easy chair and shifted some pillows and blankets from the sofa. Vasin had counted at least two other rooms leading off the hall, but evidently Axelrod still sometimes slept on his couch rather than in the bedroom. Vasin sometimes did the same himself, though in his own case it had been to escape from Vera into his own thoughts.

'What did your machine tell you?'

'It's a machine. It performs the calculations you give it.'

'So what did it tell you about the tamper?'

'I'm trying to explain. You put in variables, and it applies formulae to them and comes up with a result. Boil a liter of water at one hundred degrees, turn it into steam, and what volume does it occupy? Sixteen hundred ninety-four liters. First time you measure it, you have to do a physical experiment. But once you have the data, the second time a computer can do it for you. It's called a mathematical model. And you can scale it up. For instance, to project the effects of a nuclear explosion.'

'So why test the bombs at all?'

Axelrod looked at Vasin as though he were simple.

'You do the real-world test to show the world you can do it.'

'So what does it say about RDS-220?'

'That's what I'm trying to tell you. Nobody really knows exactly how the hydrogen – actually tritium and deuterium, heavy versions of hydrogen – will react as it undergoes fusion. You just have to go on the previous results and assume the reaction will be similar.'

'Assume?'

Axelrod shrugged. 'You assume until the experimental data proves you wrong.'

'RDS-220 is an *experiment*?'

'All new devices are experimental, by definition.'

Vasin saw no trace of self-doubt in Axelrod's boyish face.

'I've entered new data for Adamov's lead tamper. The machine ran for four hours performing a hundred thousand operations per second. So far it's less than halfway through the complete model . . .'

Axelrod made a grimace of doubt.

'Out with it, Axelrod.'

'It's just a partial picture. The reduced density of the new tamper means that the reaction is contained for a shorter period. This we knew. And the nonreactiveness of the lead means we have to subtract all our projections for the reaction of the uranium in the old tamper.'

'Meaning that the power of RDS-220 is reduced. By how much?'

'So much of this is guesswork until the model is complete, Major.'

'By how much?'

Axelrod swallowed nervously.

'My guess would be perhaps fifty percent.'

'Exactly as you said before.'

'As I suggested before, but once the computer modeling is complete we will have the data to prove it. Deliberate sabotage.'

Axelrod looked at his hands. His voice had become quiet.

'Dr Petrov confided in me,' he said. 'He often mentioned Professor Adamov's anti-Soviet attitudes. Adamov makes no secret of his subversive opinions to his intimates.

'The great Adamov also served time in a correctional facility for counterrevolutionary activity. There is a stain. Petrov told me that a time would come when such irreverent political attitudes would no longer be tolerated. That a project like RDS must be led by ideologically pure cadres.'

'Perhaps Dr Petrov considered himself a suitable replacement for Adamov, one day?'

'Adamov treats the program like his private kingdom. Petrov was passionate. He felt the Director lacked zeal. That he was too cautious. Fedya, Dr Petrov, said that Adamov was scared of the bombs he was creating.'

'And yet Adamov built RDS-220.'

'Yes, but now he is unbuilding it.'

'You think he poisoned Petrov?'

Axelrod shrugged.

'I think he is capable of murder, yes. But Adamov is undisputed master here. He did not need to *kill* Petrov to get his way. So I don't understand.'

That makes two of us, thought Vasin.

'Maria Adamova thinks that . . .'

'I don't much care what Masha Adamova thinks,' Axelrod interrupted, his meekness abruptly discarded. 'She's a lying whore.'

'"Her or me?" Those words mean anything to you, Axelrod?'

Axelrod sank back on the sofa, dislodging a sheaf of papers. He sat very still as the pages slid, one by one, onto the floor.

'She showed you?'

Vasin let the silence build. He needed Axelrod good and scared. He needed to be sure of the scientist's silence.

'Yes. She showed me the photos, Doctor.'

'What will you do?'

'I will bear them in mind, Comrade. Nothing more. Which is precisely what you will do with this information about Adamov's new design for the tamper.'

Axelrod blinked, as though the thought was crossing his mind for the first time.

'I . . . I don't understand.'

'You will talk to nobody.'

Axelrod nodded meekly.

'Nobody. Only me.'

The young scientist bit his lip. Vasin felt a sobriety, a clarifying loneliness, in the young man. He seemed to be thinking intensely about his hands. His brown eyes held Vasin's for a moment, slipped away, and came back to him as though Axelrod had reached some kind of resolution.

'You pull your hook tight, Major.'

In the courtyard the sky had turned dark with low cloud. Light snow was swirling in the wind trap made by the enclosed space. Vasin checked for the mysterious watcher, and for his own familiar *kontora* team. There was no immediate sign of either. He took up his position behind the wooden clown. There was a clatter on the stairs, and Axelrod hurried out. He had changed into a checked suit and a bulky overcoat against the chill, and under his arm he carried a slim cardboard document folder. Off to an important meeting at the Citadel, Vasin guessed. As his quarry disappeared into the archway, he slipped silently into Axelrod's wake. Stepping quickly across the pedestrian sidewalk into the shadow of the trees that lined the road, Vasin watched the scientist trotting across Builders' Street toward the tram stop on Gogol Boulevard. A familiar hunched figure broke cover from the bushes and followed Axelrod.

To Vasin's relief, there was a small crowd at the tram stop. Axelrod was anxiously checking his watch, oblivious of his suite of followers. When the tram clanked into view, Vasin muttered another prayer of thanks; it was a double-length car, made of two carriages coupled together. He waited until Axelrod, his tail, and the rest of the crowd had pressed their way on board the first carriage before darting into the closing door of the second car.

Schultz. Boris Ignatiyevich Schultz had been the name of Vasin's instructor during his three months' countersurveillance training at the Dzerzhinsky Higher School of the KGB. Little white hands, a survivor's gravity. The other trainees, all much younger than Vasin, whispered that Schultz had worked in Tokyo and Shanghai among the community of deluded dreamers, the lost rabble of prewar fellow travelers of the Communist International who had later been erased in the Purges. Somehow Schultz had returned alive. Word was he'd given up his ideologically impure comrades in exchange for his life. In person, Schultz had been dry as a closed book on a library shelf. But he was good, very good, at watching.

When they had reassembled after an afternoon on the streets around Michurinsky Prospekt trying to avoid Schultz's all-seeing eye, their instructor would greet his pupils with a slow reading from a notebook filled with times and places.

'Comrade Vasin. At 14:02 you attempted a dry-clean in the toilets of the Warsaw Cinema,' Schultz would recite in his mysteriously accented Russian. 'At 15:15 you exited the metro train at Prospekt Marxa, walked to Teatralnaya Station, doubled back through the North Corridor . . .'

The man was a limpet. Yet when he set the whole fifteen-strong class on himself, Schultz disappeared like a noonday shadow. Inevitably, they would find him silently waiting for them back in the classroom, swaying in quiet dignity after a few comforting cognacs at the cafeteria.

A few weeks' training under Schultz hadn't made Vasin a professional, at least not by his teacher's exacting criteria. But one of the details that Vasin had learned was that all Soviet trams were double-ended, with identical rearview mirrors front and back, giving a perfect view of whoever entered and left if you stood by the second leaning post behind the driver's cab. As expected, Axelrod descended at Kurchatov Square, his shadow ineptly leaping out in his wake. Vasin himself jammed the door open with his weight, ignoring the imprecations of the citizens pushing past him to get on, and stepped off just as the tram gathered speed. Axelrod took the broad steps of the Institute two at a time. The man who was following him watched him go up. Vasin joined a group of men examining the evening papers that were pasted on boards by the tram stop, in full view of the road but hidden from Axelrod's tail. He watched the *kontora* Volga spot him and circle round the square. Not Schultz-level training, but at least they weren't total idiots.

The man loitered conspicuously by the steps before checking his watch and evidently deciding that Axelrod had settled in for the afternoon. With a rolling step like that of a sailor from a children's animated film, the man began to walk down Lenin Avenue. The Volga with Vasin's own tails began to follow him more closely, evidently spooked by his suspiciously last-minute tram hopping. Perhaps they had already radioed for a backup team. The goons' boring day of surveillance was suddenly getting interesting.

Vasin had been taught to follow a target, but to do so when he was himself a target — that was a conundrum that would have intrigued even the cold-blooded Schultz. But Vasin's more immediate problem was that Sailor was heading out of the boulevards of the center into the thinly populated side streets of Arzamas.

There's only a certain number of times a subject can notice you and not realize he's being tracked. Sometimes it's just once. Best bet is on the third time. To hope for more grace is, as Schultz had put it, 'sheer optimism', pronouncing the phrase as though optimism was a foolish weakness. Vasin was pretty sure that Sailor had clocked him after he turned off Peace Prospekt into a street of low, single-story wooden dwellings not even dignified by a nameplate. The street was wide, with sidewalks of beaten earth and grassy ditches on either side of the paved surface. These were the humble places of the town, the places where the high, proud facades that would see in the future of socialism melted away to reveal the higgledy architecture of peasant Russia, not quite yet consigned to the dustbin of history. The only cover was telegraph poles. Sailor deliberately slowed his pace, giving Vasin no option but to do the same. He sauntered to the entrance of a junkyard and turned the corner. Half a second too soon to be completely out of sight, he broke into a run. Sailor was bolting.

Vasin sprinted into a yard filled with old trucks in various states of dismemberment. The place was enclosed on two sides by a high, crooked fence, and on the other by a series of barnlike garages. A door banged, and as Vasin ran to follow he saw the *kontora* Volga lumber gingerly into the yard behind him.

The service door of the garage opened into a narrow alley that snaked between the backs of wooden houses. Sailor's black figure was disappearing at speed into a small grove of birches.

'Stop!' shouted Vasin, already breathless. 'Stop!'

His only answer was the scream of a steam whistle and a cloud of drifting smoke from a passing locomotive. So they were by the railway lines. Vasin ran on, through the damp darkness of the little wood, and emerged into the shadow of a vast, hangar-like building that walled off the end of the path. Nobody to the left or right. But in the back of the hangar, an open steel door. Vasin darted into a space of darkness. He fumbled for the bolt and found it. For a second, he hesitated before shooting it home. Did he want to do this alone? He could hardly rely on the *kontora* goons to rush to his aid. The less company the better. The steel slid clumsily into place, shutting out his own pursuers.

The afternoon light, already fading, barely filtered through grime-stained windows in the roof. Vasin made out giant, dark forms around him, and smelled soot, oil, and rust. He reached forward and touched a wall of cold metal. His eyes, adjusting to the gloom, made out the giant shape of a locomotive, then another. Underfoot, his shoes crunched on gravel. An engine shed. The sounds of distant traffic echoed in the vast hall.

To his left, a pigeon fluttered up to the roof, the noise explosive over the thrumming of blood in his ears. There had been no sound of footsteps running across gravel. Sailor must still be close by, hiding. Vasin crouched by the locomotive's driving wheel, reaching up to the driving rod to steady himself. His hand closed on cold grease. He peered underneath the engine and saw only a forest of wheels and spokes, the dull steely glow of parallel rails, immobile shadows. Treading carefully, he crept to the front of the engine. The buffers thrust out like a pair of bowed human heads. From the rail yard came the sound of a steam whistle, and the voices of a work gang guffawing and swearing as they approached. Four men, dressed in drab gray overalls, carrying lunch pails in their hands and tools over their shoulders. As they entered the shed, their voices reverberated around in the cavernous space.

Somewhere ahead, from the darkest shadows behind a coal tender, Sailor made a break. He ran in a loping crouch, low to the ground like a chimpanzee, ducking under one locomotive and then another. Vasin, stumbling on the oil-soaked gravel of the railbed, scrambled to follow. The man straightened as he ran into the light and began sprinting across the weed-covered yard, handily hopping over the rails. The surprised voices of the work crew called after them.

Sailor reached the rail yard's concrete wall. A stab of cramp in his side doubled Vasin over, but he ran on. Sailor was trapped. But no, he reached a steel gantry that supported a row of signals over the main line on the far side of the wall. A set of rungs ran up the side. With an animal's athleticism, the man swung himself up the narrow ladder and began to climb. Vasin stopped, breathless, at the base of the gantry, watching the rough pigskin boots disappear over the top. Swearing under his panting breath, Vasin followed.

The top of the signal gantry was barely half a meter wide, two girders linked by haphazardly welded crossbars with a flimsy-looking handrail along one side. Sailor was clearly on home turf, shimmying sideways along the walkway with a practiced movement, oblivious of the ten-meter fall to the tracks below him.

'Stop or I shoot!' The words sounded ridiculous even as Vasin yelled them. His service Makarov was hanging in its holster on the back of a chair in his bedroom. He could never hope to catch the man on this icy steel. Vasin bowed his head in exhausted defeat, gasping and cursing every Orbita he had ever smoked. A steel box welded into the corner of the gantry caught his eye. Vasin flipped it open and saw that it contained a rusty tangle of rivets. He fished one out. It was twenty centimeters long, heavy in the hand as a half-liter vodka bottle. Sailor was more than halfway across the gantry, some thirty meters from him. More in frustration than from a desire to hurt him, Vasin pitched the rivet after the fleeing man. It ricocheted, with a ringing clang, off the steel cowling of a signal.

Sailor stopped and turned to Vasin, scowling in fear and alarm. He had a brutal, low-browed face, much older than Vasin had expected, at least fifty, he would guess, with a brawler's physique. Vasin reached down for a second rivet and pointed it at the man like a pistol.

'I said stop! State Security!'

The wind carried away Sailor's answering insult, and he continued his scramble toward the other side faster than ever. Vasin threw the second rivet, uselessly, then pitched a third, which bounced off the frame of the gantry, and a fourth, which once again dinged against a signal. With each impact the man ducked. He was almost all the way across now. Vasin fished in the box for something weightier, and found a wrench as long as his forearm. With a final, desperate throw, he flung the heavy tool. It pitchpoled through the air with a menacing swish. Damn, thought Vasin as he watched it fly. What if it actually hits the fucker?

The wrench missed. But as it scythed past Sailor's head, he flinched to avoid it and lost his footing. One boot skidded into space, and he went down on one knee. The man's hand grabbed for one of the uprights of the handrail and caught it, on his back with half his body over the edge. Kicking against the air, he scrambled to regain the walkway. An absurd thought flashed into Vasin's mind as he watched the man struggle. He had never seen anyone so absolutely alone, fighting to live, beyond help.

'I'm coming! Hold on!'

Reckless with adrenaline, Vasin stepped onto the steel walkway. Copying Sailor's shuffle, he edged as fast as he could along the gantry, hugging the signal cowlings for support as he passed them. He could see Sailor's hand sliding slowly down the handrail. The knuckles were white, and the outstretched arm was covered with the deep blue of prison tattoos. Hooking one elbow around the handrail, Vasin reached down and grabbed the man's trouser leg. Sailor stopped struggling, gathering strength. On an unspoken signal, they both heaved. Vasin hauled the man's knee onto the gantry. Sailor rolled onto his back, safe, his legs tangled with Vasin's.

Neither man spoke. A flight of starlings crowded into the gray sky above Vasin's head, swirling through each other like live smoke. He felt the thud of blood on his aching muscles, and the panting shudder of the man tumbled with him. He could smell the sour reek of Sailor's clothes, a tang of engine oil and sweat. Theirs was almost like the intimacy of spent lovers.

With a supreme effort, Vasin pulled himself upright. Sailor did the same. Side by side, facing the snaking steel railway lines and with their legs dangling from the walkway, they sat, sucking down air. Vasin looked sideways at the man whose life he had saved. He took in the vicious face, swollen with drink and rough living and with a smoldering violence in the eyes, like a dog that would bite if it dared.

'Who are you?'

There was no accusation in Vasin's question. He hadn't been expecting an answer, and didn't get one. Still he persisted.

'Who do you work for?'

'For the glory of the fucking Motherland. Same as you.'

Sailor's voice was a smoker's, like churning cement.

'What do you want?'

'How about you just fuck off out of town.'

'*Who* wants me to fuck off?'

Sailor looked Vasin up and down dismissively.

'Grown-ups, boy.'

Vasin tried to think of what to do next, but his head was filled with a singing vibration. The noise grew. It took Vasin a moment to realize that the noise was not in his mind but coming from below, a metallic hum rising from the rails. Ahead of them, a plume of white smoke rounded a corner followed by the black silhouette of a speeding train. It was approaching on the track directly under their feet.

'Are you hanging on tight, friend?'

Vasin's mind scrambled to find the man's meaning, even as he tightened his arm around the handrail post. The rumble of the approaching locomotive drowned out words. The red star on the front of the engine bore down on them, paralyzing in its power and speed.

'I said, hold on tight.' The man was shouting in his ear now. As he spoke the engine thundered beneath them and they were engulfed in a thick cloud of oily steam and coal smoke.

Vasin felt the man's sudden movement a split second before the punch landed, square in his groin. A strong hand immediately went to his chest, steadying Vasin as he doubled forward in a wave of pain. He gasped a toxic mouthful of smoke as an agony such as he had never experienced washed through his body. The hand moved to Vasin's arm, gripping it tightly as he reeled, holding him back from a fall. Then it was gone. When the steam thinned Vasin found himself alone in his throbbing world of pain.

'*Ty che?*' He heard a voice coming from the end of the gantry by the rail yard. 'What do you think you're doing?'

Vasin was too busy fighting back nausea to answer. He closed his eyes and felt the steps of the railway men ringing though the steel girders on which he sat. He allowed himself to be hauled to his feet and helped slowly back along the gantry, none of his muscles properly obeying him. A small audience had gathered at the base of the scaffolding, a gang of workingmen, a supervisor in clean overalls. Vasin saw frank incomprehension in their eyes, then looked down at himself. A sooty man in a good suit and mac, smeared in machine oil and grease, fooling about on a signal gantry. Brushing himself off once on firm ground, he nodded at them, summoning dignity.

'State Security.' Vasin's voice had become unnaturally high. He fumbled for his red ID card and flipped it open.

The men's eyes widened in appropriate awe. Vasin extended his hand to the man in the cleanest overalls, who was momentarily too stunned to take it. When he eventually did, the supervisor clasped Vasin's hand in both of his, as though greeting Party nobility.

'Thanks, Comrade Foreman.'

Over the foreman's shoulder Vasin caught sight of two portly men in suits lumbering across the yard, his *kontora* goons, finally catching up with him.

'My men. Too late, as usual.'

Obedient as an audience of schoolchildren, the railway men all looked round at the new visitors.

'Farewell, Comrades. Remain vigilant. Saboteurs are all around.'

Vasin picked his way unsteadily across the rails toward his approaching colleagues. At least the sons of bitches would be good for a lift home.

The two goons had offered no demur when Vasin followed them to their car and climbed painfully into the backseat. But they drove in the uncomfortable silence of men unused to breaking regulations. Vasin outranked both of them, of course. And he was that most exotic and possibly dangerous creature, an officer from Moscow, mysteriously free to gallivant about their secret city on a mission they would never dare to guess at. Nonetheless Vasin preferred to seed an explanation for his sudden chase rather than leave it to their own provincial imaginations.

'I saw a suspicious individual lingering by Dr Axelrod's apartment.' There was no point in trying to conceal whom he had met at lunchtime. 'Looked like an undesirable social element to me. Thought I should follow him. Bastard got away. I'll be putting it in my confidential report. And file a description with old man Zaitsev.'

The *kontora* men maintained their silence. Like all Vasin's comrades at the KGB, they had good ears for the unspoken. Above your heads, Comrades. Ask no questions and you will get told no lies. Sailor's words echoed in his head. 'Grown-ups, boy.' He'd heard that patronizing, old prisoner's tone somewhere before. Vasin settled back on the Volga's worn upholstery and concentrated on a hot shower, a change of clothes, and fighting the waves of pain that rose from his balls.

IV

Two hours later, showered and dosed with all the aspirin he could find in Kuznetsov's medicine cabinet, Vasin found the Arzamas-16 Univermag busy with late afternoon shoppers. Again, he was struck by the strange calm of the place. Unlike their Moscow counterparts, the customers strolled casually by well-lit and fully stocked storefronts, unimpressed by the bounty of the place. Vasin had always dreaded his occasional family visits to Children's World, to GUM or TsUM, the multistory emporia of the capital, where his fellow Soviet citizens approached shopping as a kind of viciously competitive blood sport. The chief prize, in Vera's view, was to obtain the very thing that everyone else wanted, even if they had entered the shop with no intention of buying Czech shoes, Cuban rum, Volga butter, or the new kind of Mikoyan sausages. The three of them would wander the wide corridors, desultorily browsing displays of fountain pens, withered apples, and galoshes until Vera, with a herd animal's instinct, would catch the first tremors of mass movement. Bananas! Sneakers! Tracksuits! She would question the waddling housewives as they passed and receive vital intelligence, flung backward in haste. 'They say they're putting out sprats on the ground floor!'

Sometimes, the rumors were baseless, though the crowd stubbornly refused to believe the assistants' denials. Vasin and Nikita had once spent two hours waiting in the women's lingerie department, in a silent agony of mutual embarrassment, for a phantom shipment of brassieres, while Vera herself went in search of tins of orange juice. Nikita was to run and get her if the goods arrived while his father joined the queue. On another occasion they had been luckier. Vasin, mortified, had ridden the trolleybus home festooned in dozens of rolls of toilet paper, strung together in bandoliers with twine. It had been the envious looks of his fellow passengers that discomfited him, not the paper's intimate purpose. But here, he found himself in a land of effortless plenty. It was as though socialism had finally arrived in a single secret city in the Middle Volga.

Today, though, Vasin ignored the artfully stacked tinned goods and scanned the place with a professional's eye. A three-story maze, perfect for slipping away from watchers. Two public staircases and a service one. A customers' lift and a goods lift. In the basement Vasin found a popular buffet crowded with customers. And around the corner, a door marked STAFF CANTEEN. He doubled back, watching for the slurred step, the eye that ducked his glance. None of his tails were visible. He paused.

To the *kontora*, this confidential communications system that Masha had proposed would look like clandestine and personal communications with a key witness. Vasin would think the same if he were in their shoes. If he contacted her secretly, he would be in her power. So – why do something so foolish? To get to the bottom of the Petrov murder by any means necessary. That was Vasin's official explanation to himself. It was a self-delusion that he clung to all the more stubbornly because he knew that was exactly what it was. How could Masha really help him? She had no access to the laboratory records of Petrov's thallium experiments, which Axelrod insisted had been faked. Zaitsev was sitting on those. Was he hoping for some dramatic confession from her? Would he ever learn what really happened at the fateful dinner with Petrov? Not a chance. She had pointed the finger squarely at Axelrod.

What did Vasin really want from her?

He felt the tang of the iodine that Masha had applied to his bruised forehead the night he found her on the roof of the Kino. God knew, he needed her healing touch now. He'd run out of leads. Axelrod had been squeezed as dry as Vasin dared. Korin hardly seemed in the mood to deliver another of his curt lectures. Adamov would only speak to him the following day, after the bomb was on its way north. Masha was the only part of the mystery he had left to explore. Vasin stooped to tie a shoelace and shuffled through the door to the staff canteen at a crouch.

Maria's man Guri was the immediately recognizable king of the sandwich bar, lord of all the sausages. The man was big and swarthy, with waved dark hair parted like a razor slash. He was broad and thick-faced and violent, with pressed-together lips and a patriarch's paunch. The Georgian wore a stained cook's apron and barked orders to the waitresses as though to junior members of his extended family. He slipped the apron over his head, wiped his hands on it, and tossed it toward the cashier's lap without looking where it landed. From a drawer he produced a sheaf of forms, bumped it shut with a swing of his haunch, and crossed the room to intercept Vasin.

'Greetings, Comrade. How is business at the Gorky Meat Processing Plant?' The Georgian had donned the fastidious grammar and ingratiating manners of a Soviet middle manager in the presence of a superior, though Vasin sensed that both were returnable without notice anytime Guri damn well felt like it. 'Your meat products are the finest in the Soviet Union! Good to see you again.'

With exaggerated politeness, he gestured Vasin to an empty table. Guri spread the order forms before his guest as though he were dealing cards before sitting down himself and leaning forward confidentially.

'Masha's friend?' he muttered.

Vasin nodded and repeated the lines as Masha instructed.

'I'm looking for Seraphim.'

'You found him. What can I do for you?'

'She mentioned you might offer a communications service.'

'Yes, esteemed Comrade,' Guri said, after a slightly eerie delay. 'For friends.'

'And how may I become your friend?'

'All of Masha's friends are friends of mine.'

'You make friends easily.'

A wide, gold-toothed smile spread across Guri's face like an opening theater curtain. His conjurer's hands spread wide, as though a white dove had just fluttered out of his order book.

'There are no strangers in this world, only friends you have not met yet. My mother taught me this, sir. We men of Gori are famous for our friendliness. We are always ready to help our fellow beings through this vale of tears, however we can. Even if it is only with the tiny trifles of life.'

Some people transmit, Vasin thought. They bring you their whole past as a natural gift, open their heart before you like a book. Some people are intimacy itself. Vasin had met plenty of the sons of bitches.

'Delighted to hear it, Comrade.'

'Will it just be communications services you are requiring today, sir? Nothing else?' Guri raised his brows suggestively.

'Communications.'

'In these disordered times, people have need of all sorts of things. Things only Guri can provide.'

'Disordered?'

'The whole town running because of that infernal machine the Golden Brains are building over there on Kurchatov Square. They say it will kill all the capitalists. It's making everyone crazy. Actually ran out of condoms this week. Had to send a man to Moscow to fetch an emergency shipment. Don't ask me why the sudden demand. When I have had a hard day's work, my dear lady wife can be sure of an undisturbed night. But these brilliant young kids. So energetic! Spend all day screwing in their nuts and bolts, and then all night . . .'

Vasin tried to banish the image of Guri, the mountainous bedroom athlete, in bed with his wife from his mind.

'Nothing easier, Comrade. But might I trouble you to complete a few formalities? You understand, of course.'

This man should be an actor, thought Vasin. His face was a cinema-screen of false emotions.

Vasin followed Guri as he waddled across the cafeteria, through a kitchen foggy with cabbage-scented steam, and into a storeroom.

In the corner was a flight of old stone steps, obviously part of an ancient building that had stood on the site before the Univermag was constructed.

'Here, please. A journey into history!'

The steps led down to a vaulted stone cellar. The walls had been plastered haphazardly, as if by drunks in a hurry. An office desk was crammed into one corner, and around the walls boxes of produce were stacked. A heavy wooden door, locked with three solid padlocks, was partially concealed by a tacked-up curtain. The place had the faint, rotten tang of a pond.

'How old is this place?'

'Who knows? You must ask someone with an education, like yourself, not poor Guri. As old as the monastery just across the river.'

Grinning, Guri hauled out a large, unlabeled bottle that could only have contained home brew. Another rummage produced a pair of cafeteria glasses.

'*Chacha*.' The grape schnapps of Georgia. 'We drink to brotherhood of peoples!'

Vasin had known Georgians, usually shortly before putting them in jail. Resistance was useless.

'A drop.'

His host made a growling noise in the back of his throat that meant: Nonsense, man. He splashed the yellowish *chacha* into the two glasses. They both drank, Vasin wincing from the fire of the home brew. Then the companionable entwining of the eyes, the tiny nod that always follows a toast. A small ritual, shared.

'Now. Business.'

Guri settled himself onto a stool and nodded Vasin to do the same. He slipped out a large green ledger and, tongue stuck out in concentration, opened it at the last page.

'Your identification please, Comrade? A necessary formality.'

Vasin hesitated for a moment before producing his scarlet KGB ID card. He kept his eyes away from Guri's. A gesture of tact, allowing the man time to find a suitable face. But when he looked back Guri was carefully copying the details, utterly unimpressed by the sword-and-shield emblem on the little scarlet folder. He returned the card to Vasin with both hands, as if to emphasize its preciousness.

'Get many of my colleagues coming to see you?'

'All my friends appreciate my discretion, friend. But Guri says, we all walk under the same sun. What matter the uniform we wear?'

The Georgian smiled broadly once more and closed the ledger. With the back of his hand he wiped his mouth each way as if he had a mustache, which he did not.

'Please forgive me, but payment here is required in advance. For friends of Masha, a small cash deposit is sufficient. At your discretion.'

Vasin rummaged in his wallet and produced a three-ruble note. The Georgian made no move to take it. Vasin replaced the bill and fished out a twenty-five in its place. As he took the banknote from his new client, Guri rubbed it, instinctively, between thumb and forefinger as though it was a piece of fine cloth. A familiar gesture, to Vasin, from his cop days. The unmistakably crisp smoothness of the government paper: no forger could ever quite match it.

'One final thing please. I think I do not need to explain, as you are obviously a man who knows the world.'

The Georgian lumbered to a corner and picked up a large cardboard box printed with garish Western lettering.

'What's that?'

'Kotex, for ladies.' They both blushed. 'From America. Arzamas is the land of plenty, but some things, only Guri can provide.'

He handed the box to Vasin.

'What am I supposed to do with this?'

'Simply hold the goods please, Comrade.'

Guri took a leather-cased camera from a hook and popped open the cover. An expensive Zenit thirty-five-millimeter, Vasin noted, military-grade. Such a camera would cost him a month's salary.

'With your permission?'

Vasin indeed understood without having to be told. These photos of customers clutching contraband were Guri's insurance policy. He stood, trying to compose his face into a suitable mask of disapproval in case the photograph ever found its way into the hands of his colleagues at the *kontora*. Guri snapped off three frames in quick succession and closed up the camera.

'It is amazing how grateful ladies can be when made a present of such a product. I can sell you a box for . . .'

'I just want to send a message.'

'Of course. Please, write down where you can be reached. And take note of our line here in the kitchen. It is always staffed. Any special words you might like to use, understandable only to your-selves. Then we do the rest. Discretion is assured. Do you have a message you wish to send now?'

'Yes. For Maria Vladimirovna Adamova.'

Guri bowed to write on a scrap of paper, his pencil poised.

'I need to see her. Tell me what time and where.'

'Nothing else?'

'Nothing else.'

'Two rubles. Including delivery of reply, of course. I may deduct it from your deposit if that is sir's wish?'

Vasin was growing weary of the man's parodic obsequiousness. Guri was a parasite, a capitalist, a speculator. He hated the complicity that the Georgian impudently presumed had grown between them.

'Yes.'

'Very good. Expect a call from the children's toys department. Or maybe electrical goods. We run a very well-stocked shop.'

As he mounted the steep stone stairs, Vasin felt Guri's appraising eyes on his back.

V

The Univermag porters fidgeted impatiently by the entrance, wait-ing for the final lingering shoppers to leave. Vasin spotted his pair of watchers, exposed by the draining-out of the crowd. They did not bother to conceal their relief when he reappeared. One wore brown, the other tweed. The brown was a swarthy forty-year-old bantamweight with scars on his left cheek. The tweed was heavier, a slab-faced man with the complexion of an overboiled dumpling. Both glared openly at Vasin as he passed. His twenty-minute disappearance had been noticed. Worse, Vasin guessed that he had forced the men to overstay their shift. In the police, he had known men to break a suspect's fingers in their impatience for a confession before dinnertime.

How long did he have before the *kontora* began to intensify its surveillance? He'd already outstayed Zaitsev's deadline. Vasin was somewhat protected from the official KGB men by the dangerous mystique of Special Cases. Masha and Axelrod were cloud dwellers, ordinarily beyond the reach of the likes of Zaitsev. But who would protect them all from the irregulars, brutes like Sailor? He felt the realm of the possible tightening around him. The goons of the local *kontora* would not dare to stop him in his duties. But soon their oppressive presence would grow so confining that he would no longer be able to move, to work.

The test was in three days. If he hadn't found out who killed Petrov by then, he probably never would – or rather, it would no longer matter. If RDS-220 was successful, Adamov, Axelrod, and Korin would all be heroes, whisked off to Moscow to be loaded with medals and then on to some luxurious Party sanatorium for a well-earned holiday. If not, they'd all end up in some extended *kontora* purgatory where Petrov would be the least of their problems. Or Arzamas and the rest of the USSR would be reduced to a prairie of radioactive dust. With a wince, Vasin quickened his step. But he could not outpace the familiar Volga that followed him down Engels Prospekt at a slow, menacing crawl.

He thought of another solitary supper at the station buffet, but felt too exhausted. As he mounted the stairs to the apartment, he prayed that Kuznetsov would not be at home. He opened the door softly, noticed his handler's overcoat lying crumpled on the floor below the coat hook, and cursed silently. He crept down the corridor without switching on the light. But within a step of his bedroom door Kuznetsov tugged open his own, throwing his stocky shadow over Vasin like a net.

'Got you!'

Vasin turned wearily.

'Busy day, actually. Didn't want to wake you.'

'It's seven thirty. It's the provinces, but even our bedtime is later than that. You need to get ready!'

Kuznetsov stood in service breeches and stockinged feet, his unruly hair plastered into place with oil.

'Earth calling Sputnik.' As if to demonstrate what he meant, Kuznetsov pulled his braces up over his shoulders with a snap of elastic.

'Ready for what?'

'Don't you have Chekists' Day in Moscow?'

Of course. The anniversary of the founding of the Soviet secret police. The red-letter day when every KGB man in the country got good and drunk. No wonder the goons at the Univermag had been impatient to get to the celebrations.

'Fuck.'

'I'm excited, too.' Kuznetsov struggled to button the top button of his collarless uniform undershirt. 'My feelings, exactly. An evening in the refined company of our comrades and brothers-in-arms. Come on. We have twenty minutes to get to the Officers' Club for the big banquet.'

Vasin sank back against his bedroom door and gently banged his head twice on the flimsy wood. Kuznetsov's big laugh was more sympathetic than mocking.

'Before you ask, yes, you have to go. The brass want to keep you close. Especially with most of the *kontora* busy filling their bellies. What would Vasin get up to if we let him out from under our eye? they're thinking. Nothing good. Just explaining, Comrade. Full dress uniform. Decorations, for those who have them. Two minutes and I'm out the door.'

—

The KGB Officers' Club was a neoclassical barn, a Stalinist parody of a porticoed manor house. Kuznetsov parked his jeep diagonally across a curb and jumped out, gesturing at Vasin to follow. Every window was brightly lit, and the sound of a brass band spilled onto the street. The main reception room was a sea of dark uniform green, interspersed with the candy-colored evening dresses of the *kontora* wives.

Vasin and Kuznetsov paused at the door.

'Christ.'

'We don't mention Him around here, friend.' Kuznetsov intercepted a white-jacketed waiter and grabbed two vodkas from a silver tray. They knocked the drinks back simultaneously and exchanged a grin of mutual sympathy. Vasin was glad of his companion's humor. He exhaled loudly, like a swimmer about to take a plunge.

'Is there a Mrs Zaitsev?'

'Oh, my word, yes. Just you wait.'

They both laughed indecorously loudly, drawing glances.

At that moment all conversation was interrupted by a majordomo flinging open the double doors to the ballroom, where a buffet would be waiting.

'Comrade Officers, dinner is served!'

It was not quite a stampede, but came close. Vasin was nearly knocked off his feet by a refrigerator-size woman who had actually hitched up her long skirts in the race for dinner. A colonel hurried in her wake, clutching their two glasses. The room emptied faster than if someone had shouted 'Fire!' By the time Vasin and Kuznetsov succeeded in squeezing their way through to the devastated buffet, most of the platters of delicacies had been picked clean. Only a tangle of severed pink claws testified to vanished Volga river crabs. The black caviar sandwiches had gone, leaving only red. Pineapple heads lolled, bodyless.

'Khan Mamai and the Golden Horde have been through,' Kuznetsov laughed. Nonetheless they loaded their plates with mushroom pies, chicken vol au vents, salami sandwiches, and piles of pink sliced ham. The drinks they had downed as they waited for the scrum to thin had warmed Vasin and sharpened his appetite. He hadn't eaten since breakfast.

'You don't do too badly,' Vasin replied, smiling through a mouthful of pie. 'For provincials.'

The band, reassembled in the corner of the ballroom, struck up a ragged fanfare. Quiet spread through the company. A group of loudly chatting subalterns, the last to notice that the official part of the evening was beginning, were shushed into silence. A line of waiters, each bearing a tray of shot glasses, filed into the room as neatly choreographed as a corps of ballet dancers. The bulky figure of General Zaitsev, drink in hand, lumbered into view on a podium at the end of the room.

'Comrades!' There was no microphone, but Zaitsev had the impressive parade-ground bellow of the Revolutionary generation, for whom such fripperies had been unnecessary to rouse the masses. 'On this day of glorious memory, we gather to honor those who have preceded us.'

Vasin felt the familiar sensation of swooning backward into himself. A conditioned reflex of every Soviet citizen, to switch off one's eyes and ears during moments of orchestrated boredom, to turn inward. A way to be alone, even in a crowd. The booze helped him think of nothing. He swam weightless in the swelling warmth of vodka and the rising and falling burble of Zaitsev's words. Only a pause in the General's monologue broke his reverie.

'But today is also a day to look forward. Those in this city who work in peace and security thanks to our efforts will soon demonstrate a device that will once and for all establish our superiority over the Imperialist enemies beyond our borders. That battle will be over. But on that day another battle will begin. A battle with the enemies *within* our nation.'

A tremor ran through the audience. Glances were exchanged, nodding listeners nudged into attention.

'Yes, Comrades. There are some in this nation who say that there are no more enemies. There are those who hold our work in the

Committee for State Security in contempt. Say that we are butchers. Mock us not just in their kitchens, but even in the highest councils of the land. To those saboteurs, to those traitors, I say: The time will come when we, the guardians of State Security, will turn to restoring ideological discipline in our own nation. And it will come soon.'

The silence was electric. Vasin struggled to believe what he had just heard. Had Zaitsev just publicly called the leaders of the nation *traitors*?

'To our heroic Soviet Motherland!'

Everyone drank a sip, by a tradition known to every man and woman in the room, ready for the customary second toast.

'And to our Service, ever vigilant in the defense of her borders and in the hunt for her enemies!'

This time the glasses were tipped all the way. Vasin felt his toes curling up inside his boots as he rocked backward, on the verge of losing balance. Steady, man.

An excited buzz of conversation broke out as Zaitsev stepped down from the podium. The band struck up an incongruously frivolous up-tempo polka, shattering the tension, and his fellow Chekists began to steer their wives onto the dance floor. With Zaitsev making his backslapping progress across the room in his direction, Vasin turned to find Kuznetsov gone. He'd missed the speech.

He found his handler near the bar huddled at a table with three fellow officers, cognac and vodka bottles arrayed in front of them.

'Ah! Our Government inspector. Vasin, come join us. Our colleagues have been dying to meet you.'

Kuznetsov clapped a strong arm around Vasin's shoulders, as though his quarry might make a break to escape, while another man whipped a chair away from a neighboring table.

'Vasin, meet Kesoyan. Oskolkov. Shubin.'

To Vasin's embarrassment, Kuznetsov's companions stood for the introductions.

'Kesoyan.'

The slight Armenian major with a fastidiously trimmed mustache gave him a comradely handshake of quite vicious respectability.

'And I'm Oskolkov, sir,' a young lieutenant privately explained, just as respectfully, peering over Kesoyan's shoulder. But Oskolkov was evidently not in the handshaking class: Kesoyan had done it for both of them.

'Shubin!'

A ruddy farmboy's face incongruous over his major's bars. Shubin shook his welcome across the table and grinned with drunken bonhomie. Amid a hospitable reshuffling of chairs, a round of drinks was poured and knocked back.

'We've been discussing the news,' said Shubin.

'Front page of the *Red Star*,' chipped in Kesoyan.

Vasin spread his hands.

'Oh, you know what Moscow is like. We're the last to hear everything. I come here to get up to speed.'

His feeble joke earned him a round of smiles.

'We launched a nuclear-armed ballistic missile from a submarine,' Shubin said proudly. 'It was fired from a Project 629 boat from underwater. Detonated perfectly over Novaya Zemlya. Imagine, launching a missile from underwater!'

Vasin's blank response prompted Shubin to lean forward to explain further.

'It's a new era, Major. Sea-launched ballistic missiles mean we can attack the enemy from beneath any ocean. Completely undetectable. The Yankees have never managed it. This will have them shitting in their cowboy hats.'

'That'll teach Kennedy to meddle in Berlin,' Kesoyan chipped in. 'Or Cuba. Confusion to the capitalists!'

They drank once more.

'Comrade Shubin is a great seaman,' added Kuznetsov. 'Takes great pride in the achievements of our glorious Soviet Navy.'

'Major Kuznetsov is teasing me, Vasin. Next thing he's going to say is that I was assigned to submarines because of seasickness. Not true. Like most things Kuznetsov says.'

'Nonsense. I have the greatest admiration for your nerves, Shubin. Couldn't do it myself. Cooped up in a steel tube with a hundred men, deep under the ocean. And a nuclear reactor just a couple of bulkheads away. And nuclear warheads too. Like being locked in a bunker, but with the radiation on the inside instead of out. You heard about that accident on that sub, K-19, where the reactor sprang a leak? The sailors nicknamed the boat Hiroshima.'

Kuznetsov, carried away with his own speech, failed to notice his companions straightening their backs at the arrival of an outsider.

'Me, I'd be out of the first hatch, take my chance swimming with the sharks.'

Behind Kuznetsov's chair, Major Efremov gently cleared his throat.

'May I join you, Comrades?'

Oskolkov was the first to jump to his feet.

'Sit, please.' Efremov placed his hands on both of Kuznetsov's shoulders, as if to hold him down, though he had shown no sign of stirring. 'Glad to see you have been making our esteemed colleague from Moscow welcome.'

Vasin looked up at Efremov's narrow inquisitor's face. He was sober; conspicuously so. Somewhere on his prudent little journey to power, Efremov had taught himself to smile. It was an underhand weapon to use on people, rather like silence on the telephone, but effective. Efremov smiled now, a thin smirk. Though he outranked nobody still seated at the table, his presence caused every man to stiffen and compose his face. Theirs was the quiet not of insolence, but of fear.

Efremov sat in the lieutenant's vacated chair, and with a sideways glance indicated that Oskolkov could get lost. The adjutant's yellowed gaze settled on Vasin with an air of informed suspicion. Vasin could still see the smirk he wore as he'd emerged from Petrov's radioactive apartment.

'Damn glad you joined us, Efremov. Drink?' Kuznetsov picked out a half-full cognac bottle from the collection that had gathered on the table.

'You know that I do not, Major. But don't let me stop you. We all know how much *you* enjoy your drink.'

With a sigh, Kuznetsov looked upward, as though seeking divine help to control his irritation. Evidently he didn't find it.

'We were just talking about Comrade Khrushchev's speech last week,' interrupted Vasin. He felt all eyes upon him. His desire to bait Efremov was uncontrollable. It was as if they had recognized each other as heirs to some ancient feud. Flushed with drink, Vasin plunged on.

'His address to the Twenty-Second Party Congress was brilliant. Don't know about you, Efremov, but I was inspired.'

'All the General Secretary's speeches are inspiring. Naturally.'

Vasin shifted on his plush chair, positioning himself for a better thrust. There was no mistaking the menace in Efremov's tone.

'You read the speech in full in *Pravda*, of course. I'm sure you have been following the reports on the Congress every day as closely as I have, Comrade Major. The Comrade General Secretary's attack on the Cult of Personality. We know he has confided such thoughts to the Party before, privately. But now he says it in public. Brilliant. In a word.'

Efremov did not reply. His stillness had become chilling.

'Cult of Personality' was Khrushchev's code for Stalinism. I am a fool, Vasin thought fleetingly, to make this fight public. But Zaitsev's words about traitors at the top of the Party had stung him. And the sight of Efremov's smug face made him reckless. He craved the bright colors that came with passion, the glorious release of it. And he wanted to humiliate the man, rub his nose in the fact that he and his thuggish boss were on the wrong side of history.

'The Comrade General Secretary personally condemned all the facilitators of the excesses of those days. He named them all: Voroshilov. Molotov. Malenkov. Bulganin. Kaganovich. All those who allowed the senseless slaughter of so many innocent comrades during the years of repression. Of course he did not need to mention our own Yezhov. Yagoda. Beria. They have already been justly executed for their crimes. The honor of the Party could not be clean, he said, without admitting the mistakes of the past. You surely agree with Comrade Khrushchev, Major?'

Vasin had named the closest allies of Stalin, and three heads of the *kontora* who had masterminded the worst of the Purges, and themselves been consumed by them. Men who had once been Zaitsev's chiefs.

Efremov took a long moment before replying.

'It seems, Comrade Vasin, that you are a man with progressive views on these matters,' he said. 'I find myself wondering if you truly belong in our service. Some among us have come to believe that maybe you aren't really one of us at all.'

Vasin looked around the table for support, or at least understanding. Kuznetsov's face was flushed, his mouth pursed as though he were about to explode. His wide eyes were silently begging Vasin to shut up. Shubin stared in simple fascination, as though Vasin had produced a frog from his mouth. Kesoyan's smile had become tight as a dog's bottom.

'My views are as progressive as the Party's. And the *kontora*'s, therefore. Naturally.'

Efremov abruptly stood. For an irrational second Vasin thought the man might be about to come round the table and strike him, but then he saw Zaitsev plowing through the crowd toward them with the unstoppable purpose of a tractor. Beside him was a figure swathed in violent pink chiffon. A woman, as squat and square-faced as the General himself, with her hair piled into the shape of a motorcycle helmet.

They all stood to attention. A memory of the lined face of the old mess sergeant who had shown him the ropes before his first drinks party at the KGB Officers' Club in Moscow flashed into Vasin's mind. 'Brother officers don't salute each other in the mess, sir. You're fucking *brothers*.' Vasin kept his hands by his sides.

Zaitsev's fleshy mouth worked as though he were chewing on something unpleasant as his disapproving eye flicked from one man to another before settling back on Vasin. Mrs Zaitsev's face mirrored her husband's sour disappointment with humanity in general, and these specimens in particular. Her lip curled as she took in Vasin's bruised eye, his crumpled uniform and grubby collar, eyeing him like a delinquent son dragged home from the drunk tank.

'*S prazdnikom*,' the General grunted. 'Congratulations on the holiday.'

They returned the greeting in absurd unison.

Zaitsev swayed a little as he stood squarely before them, his powerful arms loose by his sides like a boxer's. Belatedly, Vasin realized that the man was drunk. As some men may be seen to be in love, so Zaitsev seemed to be possessed by a deep and awesome hatred. For him.

'Vasin.'

He spat the word to rhyme with 'fuck you'.

'Sir?'

'Find somewhere else to chase around rail yards and jump on signal gantries,' Zaitsev growled. Rudely, he had used the familiar form of address. 'What the devil were you doing?'

'Following my Socialist duty, sir. At speed.'

Zaitsev was an enemy to jokes, feeling an energy in them beyond his control. Fury rose on his mottled cheeks like spreading ink. Vasin hastily continued.

'I was pursuing a suspicious character who threatened one of your witnesses, sir.'

Drink had rolled Zaitsev's syntax back into the South Russian farmyards of his youth.

'My office has complied with all your idiot demands. Stop wasting our time.'

Vasin made his drink-addled brain focus. Perhaps Zaitsev wanted him gone so desperately that he would give up some evidence. The *kontora*'s painstaking digest of Petrov's laboratory reports – the ones Axelrod insisted had been doctored.

'As soon as I have all I require, General. The laboratory report, specifically. From our deceased comrade's workstation. That is the evidence I am missing.'

Zaitsev's eyes narrowed suspiciously.

'Report to my office. Tomorrow.'

The General turned unsteadily, and his wife shot a parting glance of outrage, as though Vasin had somehow insulted them.

Kuznetsov was furiously silent as they wove unsteadily across the empty city. He sat hunched forward over the wheel, wrestling it like an enemy. His face lit and went out again under the sodium street lamps.

'Don't be angry with me, friend.'

Kuznetsov sighed demonstratively. The elation of the drink and the unaccustomed liberation that Vasin had felt while provoking Efremov remained like a glow inside him.

'So why are you snorting like a bull? Did I say the wrong thing?'

'Where do I start?'

'That's a start. Human words, better than farmyard noises.'

Kuznetsov turned the wheel violently, throwing Vasin against the door. He accelerated to dangerous speed down the final stretch of boulevard to his apartment and braked wildly to a halt, almost demolishing an innocent apple tree. Vasin wished someone had thought of installing some kind of restraining belts in cars, like on airplanes.

'Do you think nobody else has a mind of their own except for you? You come in here, jerk the *kontora* around, pull Efremov's tail, mock every damn thing you like. Telling him about Yezhov and Beria? That's the new Party line, so I say what I like to whomever I like?'

Kuznetsov's face was livid in the lights of the dashboard.

Vasin, to his shock, saw a tear spill down his handler's face. He was no longer angry, but distraught.

'Old man, I'm so sorry if I . . .'

'You realize that your freedom comes at a price? Every word you say means a word someone else cannot say. For you, it's just some kind of fucking game. You screw with Zaitsev, with Efremov, he's going to screw with someone, anyone but you. It's time for you to go home, Vasin. You don't fucking live here.'

Kuznetsov flung open the door, then slammed it behind him so hard that it sprang back from its latch.

Vasin followed him inside in shamed silence. He caught up with his roommate as he struggled with the door lock. Inside the apartment, the phone was ringing. Kuznetsov dived inside and snatched up the receiver at the seventh ring.

He turned to Vasin accusingly.

'It's for you.'

Vasin took the telephone and turned his back on Kuznetsov. Who knew his number here?

'Hello?'

A chirpy female voice with a strong Central Asian accent announced that the comrade's order of an electric train set had arrived and could be picked up at the toy counter of the Univermag at ten tomorrow morning.

CHAPTER NINE

SATURDAY, 28 OCTOBER 1961

TWO DAYS BEFORE THE TEST

I

Over breakfast Kuznetsov was torpid and hungover, his usual bon-homie strained. He gestured wordlessly to a pot of oats he was stirring on the stove when Vasin staggered into view in the kitchen door, spooned the porridge into two bowls on his guest's nod, then busied himself with making tea.

'I'll take the tram to the *kontora* if you've got somewhere to be.'

Kuznetsov stood, straightening with difficulty as though he bore the weight of the world on his shoulders.

'No, Comrade. You're no trouble at all.'

After a courteous pantomime in front of the shower, Kuznetsov finally agreed to go in first. As he listened to the water gushing, Vasin realized how badly he needed a friend, even if no friend in this strange place could ever be a confidant.

Tugging their crumpled and cigarette-scented uniforms into place, side by side in front of the hall mirror, Vasin and Kuznetsov exchanged an involuntary frown that restored some of the previous evening's complicity. They both looked like they had been dragged through a hedge.

'To the dragon's den, Comrade?'

'If I must.'

'Oh ho-ho. You fucking must.'

—

As Vasin made his way upward through the levels of the *kontora*'s headquarters, he sensed heads turning in his wake. His notoriety in this little world was spreading. So *that's* the guy from Moscow. Vasin walked as tall as the lingering pain in his groin allowed and presented himself to Zaitsev's bleary-eyed secretary. He had a vague memory of the woman whirling in a drunken dance in the arms of some broad-shouldered ape the night before. All the resentment the woman felt for the rigors of the coming workday was distilled in her voice.

'Yes?'

'General Zaitsev asked for me.'

The secretary nodded him into a chair and frowned, evidently considering whether this disheveled man was of a rank to be offered tea. She probably needed one herself, so didn't bother asking. Returning from the kitchen with two cups, she put one in front of Vasin and sipped her own noisily. Around them, the headquarters ground into reluctant life. Trolleys of documents rumbled past the door. Gossiping voices reached them from the smoking area of the stairwell. In an unbidden impulse of solidarity, the secretary slid open a desk drawer, fished out a plastic bag of sugared orange slices, and held it out to Vasin, drawing his attention with a brief shake. They sat on, wrapped in their separate silences.

Zaitsev's arrival was heralded by a heavy stamping and puffing, like the appearance of a troll in a kindergarten play. But when the General strode in, Vasin saw that his usually ruddy face was grayish, creased with worry lines and the assorted tooth marks of age and booze. A monstrous hangover had not mellowed his habitual fury with the world. Efremov followed, irritatingly crisp and energetic. The secretary, casting a pitying glance at Vasin, scooped up the day's files and followed her boss into his office.

The electric clock on the wall ticked past 0900. Masha would be waiting for him at the Univermag in an hour. When he thought of Masha, Vasin recognized the same vertiginous feeling he had once had with Katya Orlova, a sense of inexorability, of being drawn into something disastrous by a force stronger than his reason. Except that with Katya the force had been black and nihilistic. With Masha, Vasin felt a deep and genuine impulse of protection. She needed someone to save her – or so she had said. But save her from what? Her lover, Petrov? He was dead. From her trapped life of privilege? From Adamov, the loveless old man to whom she was bound in ways he struggled to understand? Just what he could do to save her, Vasin could not say. But he also knew that he needed her. Masha was hiding something crucial from him. Vasin felt it. She knows something. She fucking *knows*.

'Comrade Major?' The secretary replaced the handset on her desk. 'The General will see you now.'

Zaitsev's office was bigger than Vasin remembered it, large enough for a regimental dance. The General himself hunched at one end of the long conference table like a plump rat on a raft, watching his visitor cross the acre of parquet with bloodshot eyes. Gesturing Vasin to sit, Zaitsev crammed back the sleeve of his tunic and twisted his wrist around as if it were someone else's. The dial of an old steel watch returned his stare.

'We have reached an end to your antics, Major Vasin. Efremov?'

The adjutant, with the mocking formality of a waiter, handed Vasin a fat gray file.

'The summary of the laboratory records,' said Efremov flatly. He paused to amplify his disapproval of the hours wasted at the visitor's caprice. 'Now you have everything.'

Vasin eyed his two colleagues warily, unwilling to believe that they had decided to grant his request. Though of course they had not spoken to Axelrod. The *kontora* men had no idea that the thallium Petrov had supposedly signed out and used to kill himself would have disappeared of its own accord through the black magic of radioactive decay.

Vasin struggled to keep the triumph out of his face.

He began leafing through the thick stack of papers he had glimpsed on the first day. Each contained a long list of dates, names, chemicals, and quantities. The cover noted that this was the fourth copy of four.

'You will note that all the reagents Comrade Petrov signed for are underlined in lead pencil,' continued Efremov. 'The thallium is in red. Each notation is cross-referenced with the experiments undertaken, including how much thallium was actually used. Over three weeks, the Comrade Doctor signed for some two thousand milligrams more than appear to have been used in the tests undertaken by his laboratory. The evidence is clear, Comrade Major. Petrov signed for and stole a classified element, and poisoned himself.'

Vasin's mind raced.

'Thanks. But why . . .'

'Why have we decided to give you the file?' Efremov exchanged a private glance with his boss before continuing. 'Because of your unique talent for disruption. Your breathtaking insubordination. You are a loose cannon, Vasin, and your activities will no longer be tolerated. Therefore – you win. You have everything. And now, you must go.'

A rumble escaped from Zaitsev, like a rockfall inside his capacious chest.

'Before we waste any more time on you, these findings clearly confirm that Petrov's death was suicide. Very clearly. My final report will be ready this afternoon. I expect you to endorse it. There's a train to Gorky leaving at 1800. And you will be on it.'

Vasin leafed on to the end of the *kontora*'s summary of the laboratory data, considering his next move.

'I can sign your report right now, General.'

'Good.'

Zaitsev pushed back his chair and stood. A thought crossed his mind.

'And what of *your* report, Major Vasin?'

'I am afraid that my investigation still has some unanswered questions. Suicide comes in many forms, General. But you have my word. I will file my final report via the proper channels—'

'Fuck your mother in her channels, Vasin.' Zaitsev's fist thudded on the table. Not for the first time, Vasin was grateful that he was not facing this man in a soundproof interrogation cell.

Vasin got to his feet and stood to attention.

'Do I have permission to leave, Comrade General?'

Efremov smoothly approached Vasin and stood uncomfortably close, his lanky height towering over him.

'You haven't understood, Comrade. If you agree to leave tonight, we give you the report at the station. The moment your boots are off the platform, I'll hand it to you. You have our word.'

Vasin looked from Efremov to the General and back again. Both had smirks smeared across their faces.

'Understood.'

'Good.' Zaitsev's mouth was curled in a wet smile of satisfaction at having outplayed his adversary.

'But I cannot be on tonight's train.'

'Why the fuck not?' The General's smile had abruptly withered into a flabby scowl.

'Because Professor Adamov himself has requested an interview with me at ten this evening, after the device has been readied for transportation. But I need the report now.'

Vasin knew that without the *kontora*'s painstaking cross-referencing he would have no chance of checking it against the Institute's original records. He prayed that Axelrod was right. If the records had been doctored, Vasin would have the proof he needed that Petrov was indeed murdered. And that could buy him some more time to find out who really had done it. If Axelrod was wrong, he would be out of time, and out of town.

'Categorically denied.'

'General Zaitsev, I plan to show Professor Adamov your findings this evening so that he can confirm your assessment. And then the case will be closed.'

'Efremov can bring it to your rendezvous with the Professor.'

'No. We will meet alone. The Professor insists. I will take the file with me now.'

Zaitsev opened and closed his giant, pistol-burned fists as though warming up for a fight. His pinkish eyes became a little redder and more fixed, those of a man sighting a natural enemy. Vasin met his gaze. I know you too.

'You're weak and subversive, Vasin,' snapped Zaitsev. 'Hid from the war. Got yourself a fancy education in Moscow. Had privileges. You may impress Adamov. You may even impress your man Orlov. I see you for what you are. You're anti-Soviet. I can smell it on you.'

Vasin stiffened at the mention of Orlov's name, but swallowed the insult. Did Zaitsev know about Katya? Vasin guessed not. The General would have blurted it out long before if he'd known. No, Zaitsev wasn't a man to play a long game. He would grab every weapon available and swing them wildly. But right now he was standing his ground, immovable as a bull. Perhaps the stubborn old bruiser could be coaxed. With a supreme effort, Vasin willed down his anger and desperation and answered calmly.

'You have accorded me much courtesy here in Arzamas, General.' Vasin kept his eyes firmly on Zaitsev; he knew that if he looked at Efremov his voice would crack. 'I will report truthfully that your investigation has been admirable and thorough and that you have cooperated fully. This will be recognized in Moscow at the highest levels. Why spoil all of this now, at the eleventh hour? You have agreed to give me the report. So do it, now. That is my final request. You have my word that I will leave for Moscow in the morning, and you will have the gratitude of all those who sent me.'

The General looked at Vasin more narrowly, as though weighing up his recognition in Moscow, his promotion, his decorations. Efremov cleared his throat, about to object, but his boss spoke over him.

'If you're lying to me, Vasin, God help you.'

Vasin saluted smartly. As he turned to leave, he cradled the papers to his chest. And once he reached the corridor Vasin broke into a run.

II

The smooth soles of his little-worn uniform boots slipped on the slushy sidewalk as he hurried along Peace Prospekt. A board in the Univermag's lobby announced that the toy department was on the fourth floor. He stopped a closing lift door with his boot and pushed his way into the crowded car, catching sight of his *kontora* tails panting in his wake. Good. Children's toys would be the last department they would think to check.

Masha had ditched her electric-blue mackintosh. He saw her dressed in a dark head scarf, crouching above a baby carriage and cooing over an infant. Vasin joined a scattering of men in the crowd of browsing young women and placed himself in Masha's line of vision, pretending to admire some tin soldiers. She waved to the baby and walked briskly in the direction of the lifts. He followed. Vasin passed Masha in the lift vestibule, grabbed her arm, and swung her through the double doors that led to the stairwell.

'Through here. My fat boys won't be taking the stairs.' He retreated to the top of the staircase, where he could observe the lift lobby through glass panels in the doors. He placed Masha in front of him as a barrier. 'Stand there. I'm keeping watch over your shoul der. Tell me where we're going.'

If Masha was alarmed by his evasion tactics, she made no sign.

'The basement.'

Vasin made a quick calculation. A uniformed man and a pretty girl would turn too many heads on the stairs. They'd take the elevator.

'Wait until one of them comes out.'

On cue, the lift doors opened and a heavyset man elbowed his way ahead of a gaggle of Arzamas matrons and headed toward the men's clothes department. What was it about some *kontora* goons that made them as conspicuous as clowns in makeup? Pulling Masha behind him, Vasin darted into the downward-bound elevator car.

They stood in opposite corners, avoiding each other's eyes like cats, as the shoppers filed in and out on each floor.

In the basement corridor Masha took charge. Instead of turning toward the cafeteria, she banged through a pair of service doors marked NO ENTRY. She led him past a bank of industrial freezers and through a storeroom. No store clerks were in evidence. Presumably Guri's doing.

Producing a key, Masha opened a steel door in the far corner and flicked a light switch. An iron staircase led down into another catacomb. There was a reek of damp, and of coal. Vasin closed and bolted the door before following her disappearing back down the steps. They reached an old stone vault, similar to Guri's subterranean private office, and also stacked to the ceiling with more boxes of contraband. In one corner was an electric heater and a cot made up neatly, Army style.

Without speaking, Masha folded Vasin in a tight embrace. Vasin allowed himself to enjoy the feel of her in his arms before breaking off.

'Wait. I need to talk to you.'

Masha stepped away from him. Her eyes were green-gray and lucid and seemed, to him, dangerously innocent. A militant simplicity gazed out from them upon a complicated world. What it is to be untamed, he thought. She feels, therefore she is.

'More secrets, Sasha?'

'No. No more secrets.'

'Quite sure?'

Vasin hesitated. The lab report was stuffed conspicuously into his mackintosh pocket. Masha followed his involuntary downward glance, then met his eyes again. She smiled tightly, a twist in the center of her face.

'Fine. You have no more secrets. Thank God.'

'Masha, I . . .'

'Please, keep them to yourself. So you wanted to see me about electric train sets, apparently.'

They both smiled at the same time, in the same way.

'I like the casual look.' Her voice was teasing. 'You took time out of your busy day to meet. You've been missing me.'

Vasin looked down at his unpolished leather pistol holder, creased tunic, mud-splashed boots.

Masha turned and bounced a couple of times on the bed, testing the springs.

'You said you trusted Guri. How do you know he doesn't have this place bugged?'

'No wires.'

She nodded toward the iron staircase and the bare vault. A single, antique electrical cord snaked down the unplastered stone wall to a metal lamp. Masha was right – not a modern, plastic wire in sight, and nowhere to hide one.

'Smart girl.'

'Heard it said.'

'About your last dinner with Petrov. I want to know what his mood was. Tell me what happened.'

The teasing tone of Masha's voice had become strained.

'Korin arrived first, grim, as ever. Then Adamov and Fyodor came together. They were exhausted too. Everyone's always exhausted around here. We had made potato soup, and the men talked bombs. Like always.'

'Do you remember if they mentioned the design of the tamper?'

Something behind Masha's face closed like a trap, but she kept her voice deliberately light.

'They talk of nothing else but uranium and tampers and deuterium and lithium-hexa-something. If I understood what they were talking about, I would make the best spy in the world.'

'So they didn't talk about a uranium tamper?'

'You're starting to be insufferable. They talked about what they always talk about, including tampers.'

'They didn't argue?'

'They always argued.'

'Angrily?'

'Korin's always angry about something. Adamov never is. Nor was Petrov. Too cool to lose his temper about anything, him.'

'Did you speak to Petrov, privately? About Axelrod, the photos you showed me?'

Finally Masha's composure broke.

'What would I have said to him? "You're a foul pervert who has no right to breathe God's good air"?'

Masha's sudden anger stopped Vasin's questions. Masha, too, seemed discomfited by her own loss of control. She stood and began to examine the boxes on the shelves with apparent fascination.

'Oooh look. This is good. I've just run out. Guri's got a new shipment of Kot—'

'I know that type. Your Guri. Wouldn't trust him.'

'Know the type, do you?' An edge of mockery had crept into her voice.

'Put plenty of them behind bars. Jolly Georgians. Be careful.'

'You live with wolves, you learn to howl like a wolf.'

'You lived with wolves?'

'Sure. Maybe I'll tell you about it one day. And I know how to deal with men like Guri. He's a kitten, not a wolf.'

'Until he needs to save his own skin. Then you see the fangs.'

'You talk to me like I'm a kid.'

'Listen, Masha, when you've seen what I've seen . . .'

Masha silenced him with a hard little fist that smacked into his arm with surprising force.

'Don't know me very well, do you, buster? I've seen some things too, Vasin.'

She was standing very close. He took hold of her upper arms. Through the thick material of her winter coat Vasin could feel how skinny she was, her limbs fragile as a sparrow's.

'Sorry. I know you have.'

Masha twisted away from his grip. Her eyes lit with a spark of defiance that he had seen before.

'You keep asking me about that damn dinner, but you want to know a real secret? *My* secret? I killed a man once. Yes. I fucking did.

It was in Leningrad. During the siege. We lived by wolves' laws. Hunting for food wherever we could find it. Running with a gang. Orphans like me ran in packs. I was a kid, fourteen, and I got caught by an air-raid warning in alien territory while scavenging. Early morning. I'd scrambled into a shelter in a cellar on Pushkin Street. People would take everything with them during the air raids, and some of them croaked down there. They were good places to look for grub. Then, just my luck, it was an actual air raid. There was nobody left fit in the city to crank a hand siren by that time. But some *Kommandatura* had an electric one, and it started singing out like a bitch.'

Masha's tone and vocabulary had unconsciously slipped into the rhythms of the Leningrad streets. She sat back down on the bed to continue the story.

'The shelter starts filling up. People stagger in, looking like corpses. Unlucky for me. I would have preferred to find real corpses with ration books in their fucking pockets. We sit there in the piss-stinking dark, listening to some poor idiots catching some heavy shit over my way, toward Moscow Station. I start to think of my crew, our food stash. Then a man comes down into the shelter, fast. The ring of good boots on the steps. *Ogo*, I think. Someone's doing well for themselves. He lights a candle stub, sniffs, looks around. A well-fed mug, padded cotton coat. Not an officer, then. A criminal. Looked around like he was some kind of underworld king. Maybe he was, but I never saw one without a pack of mongrels looking out for his back. He was alone. He takes one look at me and steps right over. Puts the candle down on the floor. Kicks my feet out from under me and lays me down. Starts rifling my clothes for food, I think at first. No one in the shelter says anything. They are in this shitty place, but their minds are elsewhere. He straddles me and starts yanking off my boots. Son of a bitch wants my boots, I think. Then his hands are pulling down my trousers. He gets one leg of my britches off. Now I know what he wants. The knife I carry in my boot clatters on the floor. While he's busy getting his dick out, I find the steel in the shadows. I wait for him to slump down on top of me, then slip it under his ribs. Before he manages to slip anything into me, in case you're wondering. Blade goes in beautiful smooth. I hold it fast while he thrashes about. His hot blood spills onto my bare belly. Man, that guy could swear . . . When I finally get out from under him, I find three watches and half a kilo of sugar in his pockets. A good bayonet. A handful of manual workers' ration books, with ID cards. They were worth two hundred and fifty grams of bread a day, each. The man was a treasure trove. Course, the other vultures in the cellar choose that moment to pay attention, start closing in. I had to threaten to cut them to keep 'em away and got the hell out of there. We lived on that haul pretty much to the end of the siege. Maybe I should be grateful to that rapist piece of shit.'

Masha had become a person Vasin did not recognize. Her voice was shrill and flinty, her face hard. When she looked up at him he saw steel in her, an open ferocity.

'Don't you dare tell me I can't look after myself.'

'I saw that on the roof of the Kino.'

Masha flushed with anger. Vasin backtracked.

'I mean – of course you can look after yourself. But I wanted to say that I've seen . . . madness. My sister, like I told you. It's more powerful than any person's will. You mustn't be ashamed. You have it in you. And you know it. God knows, it's understandable after what you've—'

'I'm not crazy.'

'If you say so.'

A tense, silent space opened between them. Masha stood, shaking her whole body like a wet dog and reassembling her image as a Party housewife. The coiffed hair, the nicely pronounced vowels. But Vasin could not shake the image of the young Masha, a wet rat in a cellar, sliding her knife into a man's chest. The blade going in 'beautiful smooth'.

Masha put her arms around him. One hand closed over the report in his pocket.

'Time to trade,' she whispered in his ear. 'I told you my secret. Ready to tell me yours?'

Vasin's hand clasped over hers, preventing Masha from pulling out the roll of paper. Her smile spread unnaturally wide as she refused to relinquish her hold on the document.

'Fair's fair. You can tell me. You can.'

She had begun the sentence in a tone of girlish cajoling – but by the time she reached the last syllable her voice had become hard.

'No.'

Vasin untangled her arms from around him.

'I have to go. My entourage will be looking for me.'

He turned toward the staircase.

'Vasin. Sasha. Wait.'

He found himself incapable of resisting the pleading in her voice. The vicious, foulmouthed version of Masha had disappeared, leaving a vulnerable young woman.

'Sasha. Tell me. Am I really crazy?'

'We'll talk it over later.'

Vasin tried to make his voice consoling. But he was thinking of the earlier flash of fire in Masha's eyes, the hard grip of her hand on the document.

She stepped forward to embrace him once more, but he was already moving toward the stairs and out of the door. He closed it behind him without looking back. He took the stairs two at a time and strode out onto the street without bothering to check if his tails were behind him.

Vasin had learned to be wary of opponents of good instinct. Now he found himself wary of Masha.

Even outside in the chilly air, he could feel the warmth of her body under his coat.

III

Vasin was a plague carrier. The realization came as he walked down the gray, chilly boulevard. His presence contaminated everyone around him. Everything he had ever touched, everything that he had ever tried to do, turned to shit. Right now he needed someone to tell him just that. He needed to speak to Vera.

The Arzamas Central Post Office, deserted at night, was by day a teeming circus of human life. Harassed housewives cradled bundles of parcels. Elderly men, their jackets sagging with medals, took their time explaining themselves to the desk clerks. Pretty young secretaries, pleased to be released from their offices on an errand for the boss, gossiped with their girlfriends. Above all, the place was a showcase of queues. The various windows answered various needs: parcel dispatch, regular post, telegrams, general delivery, bills and collections. The line for long-distance phone calls had the most restless and unsettled look. Vasin joined it behind a woman who clutched her address book open at the page where her number was written, ready for the moment twenty minutes hence when the clerk would ask for it. When the line moved, she sighed and wearily shuffled forward a single step.

The wait would be worth it, Vasin told himself. He needed to speak to Vera when she was sober. Maybe he could even head off her vengeance. Nonetheless he found his index finger jiggling involuntarily, as through practicing to cut off the call as soon as talk turned to Katya Orlova.

His forms duly submitted, corrected, and submitted again, Vasin settled down to wait for his call to be put through on a bench beside a pair of sleeping twin schoolgirls. In front of him stood the row of handsomely polished oak phone booths, ominous as a line of dentist's chairs. In one of them, soon, some small but agonizing part of his self-esteem would be unceremoniously ripped out.

'Comrade Major Vasin! Booth three!'

Again. This booth had something against him. He listened to the usual cascade of clicks and voices as his call was connected.

'Who is this?'

He had reached her at work; he could hear the shrill voices of her workmates in the background.

'It's Sasha. How is Nikita?'

'Normal. Everything is normal.'

'And how are you?'

'I'm also normal.'

'Listen, I've only got three minutes. I wanted to say . . .'

'I've packed your things. When you get back we will file for divorce. I'm not interested in your apologies anymore.'

'Please, Vera, I am sorry. But it's more complicated than you think.'

Vera interrupted him, her voice beginning to break.

'Self-righteous to the bitter end. I know you, Sasha. You use the people around you to prove to yourself how superior you are, but you're a liar. A hypocrite. I've sent a complaint to my Party committee about your affair. You're finished, my dearest. It's time you face the consequences.'

The line clicked dead. Vasin's finger had come down hard on the cradle. He held it down as though squashing the life out of a tiny enemy. He continued to hold it until his knuckle went white, as though the pressure would hold his secret back from the dangerous world listening in.

He emerged from the booth like a boxer staggering from the ring. Now it was only a matter of time before the matter worked its way through the sluggish bureaucracy of Vera's pathetic Party committee to the ears of the *kontora*. And to Orlov himself. The fatal blow was coming. He was finished.

'That *bitch*,' he hissed to himself.

He staggered out of the portals of the post office. Self-loathing replaced his fury, piercing his thin mackintosh like a cold hand. If he were honest with himself, part of Vasin would have welcomed punishment. Vera's contempt was nothing more than he deserved. The contempt of her mother and her idiot friends, too, would have been fitting. The female sex, turning their collective backs on him. As they should. But to have his betrayal, his weakness exposed to Orlov – receive his deserts from *him*. That was too much to bear. To find himself guilty before men whose own guilt was so much more profound, and infinitely more vicious. That offended his sense . . . of what? Of fairness?

'Justice is mine, sayeth the Lord,' Orlov had once told Vasin in the sarcastic, mincing tone he reserved for biblical quotations. 'None of us may choose the manner in which we meet justice, Vasin.'

Screwing the boss's wife. Not that Orlov screwed her himself, by Katya's own account. But this would be a matter altogether more serious than screwing. It would be about property and propriety, hierarchy and respect. Vasin had violated them all. He shuddered to think of how Orlov would choose to repay him.

With a violent effort of will, Vasin tore his thoughts away from Vera, Nikita, and the catastrophe that was unfolding back at the place that until this moment he had called home.

The death of Fyodor Petrov. Vasin guessed that the only thing Orlov cared about more than his own pride was his power over other men. Powerful men.

And finding a murderer in the Petrov case, serving up a guilty name for Orlov to lock away in his safe could, possibly, give Vasin a glimmer of hope.

IV

From a phone box on Lenin Square, Vasin dialed a list of numbers he had copied out from the Institute's phone directory. He eventually tracked down Axelrod at the calutron lab.

'It's your friend from Moscow. We need to meet. At your place of work, perhaps?'

The Citadel, Vasin's only almost-safe haven. A pause.

'Come to the accounts department on the ground floor. Room 109. Quarter past six.'

That gave him the whole afternoon to lull his tails into boredom, then perhaps try to lose them. And to change. As evening fell Vasin, glad to be back in anonymously civilian clothes again, lingered by the street display of newspapers, glancing around him to assess how much manpower Zaitsev had assigned to him. He saw nobody obvious. Which meant a truly enormous team.

Vasin waited for three trams of commuters to come and go, looking out for fellow loiterers. There was no sign of Sailor from the yards, at least, or anyone like him. Only a handful of grim-suited labor heroes and war veterans, their breastplates of medals jingling in the twilight, scanned the papers for want of something better to do. At six o'clock a crowd of women, chirruping like a flock of starlings as they adjusted head scarves and exchanged weekend plans, poured from the doors of the Institute. Vasin waited until the steps had cleared and hurried inside.

Axelrod had chosen wisely. The rest of the Citadel hummed with busy activity, but the accounts department was empty. Vasin found room 109, but no light showed through the window above the door. He tested the handle and found a desk-filled space darkening in the twilight. There was a strong smell of mixed women's perfumes, overlaid with the sticky odors of glue and ink. During the few seconds it took Vasin's eyes to adjust to the gloom, he thought he was alone. But then he made out the solid shadow of Axelrod's back, perched stock-still against a desk and watching the light drain from black trees outside the window.

Vasin crossed the room. He pushed aside a slide rule and an electric calculating machine the size of a shoe box to make space for himself on the desk alongside the scientist. Axelrod had hardened his face like someone preparing to take a lashing.

'What now?'

'I have the summary of the audit that the *kontora* made of the reagent records detailing every milligram of thallium Petrov removed from the lab.'

'Quick. Let me see it!'

Vasin's eyes flickered from Axelrod's face to his hands, then back to his face, checking the telltale places. Either Axelrod was a brilliant liar, or he was truly excited. He seemed to truly believe that the records would prove his lover had not taken his own life.

Vasin pulled the transcripts from the lab inventory out of his coat pocket and flourished them in front of Axelrod's face.

'You have the experiment file reference numbers? If you don't, it'll take the two of us weeks to track them all down.'

Axelrod opened the document to a random page and ran his finger down the left-hand column.

'Thank God.'

Vasin could barely keep up as Axelrod flew down the stairs into the basement. By the time they reached Laboratory Zh-4, home of the calutron, both were panting. But Axelrod hurried past the double doors of his own lab and continued down the twisting underground corridor before swinging left into a door marked REGISTRY. A young clerk, pimply and bespectacled, looked up in surprise from a thick textbook. Axelrod fumbled for a purple-striped identity card, which evidently indicated sufficient seniority to bring the clerk to his feet.

'I need file request forms. Lots.'

Axelrod's demeanor had become almost commanding. He snatched the sheaf of blank forms from the startled boy and led Vasin to a library table.

'Read out every entry for thallium under Petrov's name. Date, then experiment reference number, then reagent batch.'

Vasin began obediently reading off the ledger numbers on Zaitsev's inventory as Axelrod filled in the forms, his pencil scurrying impatiently across the paper. When they were finished he called for the clerk, who scurried over at a run.

'We need to see these inventories. Now.'

'All of them, sir? Could take a while.'

'All of them.'

Vasin and Axelrod walked to the smoking area at the bottom of the stairwell. Axelrod's fingers fumbled to strike a match, so Vasin lit his French cigarette for him. Neither spoke as they smoked. When he was done Axelrod crushed his cigarette out viciously, holding it down long after it had ceased to struggle. The reagent inventory for a single month of Laboratory Zh-4's work occupied nearly forty volumes, which the clerk wheeled out from the stacks on a trolley. Without Zaitsev's painstaking report, finding the discrepancies would have been like looking for a lost coin on a stony beach. But now Axelrod had the exact references and quickly found the relevant entries. Experiment by experiment, he and Vasin compared the quantities taken out with the amount used, recorded in the lab technician's careful hand – how much thallium was used, how much lost, how much returned.

They both saw it at the same moment. A crude enough forgery, an entry for 300 milligrams turned into 800. Vasin carried the volume to a desk lamp and raked light obliquely onto the page. There was no doubt. The page had been written over. They found another in an entry two days later, where 100 became 400. And again and again. In total, two thousand supposedly missing milligrams of thallium, falsely logged and transcribed by Zaitsev's team.

'Your colleagues have doctored the record,' breathed Axelrod. 'We need to get a citizen to witness this. Comrade!'

The registry clerk approached them, thoroughly alarmed by Axelrod's urgency.

'Wait,' hissed Vasin. Then to the clerk, 'Can we see who else signed out these records? Over the last two weeks?'

The cards were soon found in the index. Axelrod scanned the list. His own name, the ink barely dry, was the most recent. Before that, for four days in succession just before Vasin's arrival in Arzamas, a KGB Lieutenant Girkin, evidently Zaitsev's man. And before that, just a day after Petrov's death, another person accessed the Zh-4 records.

Korin, P. A.

'Pavel Korin,' Axelrod said. 'Professor Adamov's oldest comrade and fellow jailbird.'

'You're very well informed.'

'Korin's a bomb engineer. Payloads, detonators, altimeters are his area of expertise. There's no legitimate reason for him to be rooting around the experimental records of the calutron. Korin must have forged the thallium records.'

To Vasin's discomfort, Axelrod's logic was racing ahead as fast as his own.

'Korin and Adamov, they killed my . . . They killed Petrov, and now they are covering it up. I told you before, they are saboteurs. They had to get rid of Petrov because they would not be able to change the design of the bomb with him alive. This proves it. This is *treachery*.'

Axelrod's voice was becoming high and hysterical.

'Wait. Pull yourself together, man. Think. What exactly do you plan to do with this theory of yours?'

'I'm going to lodge my concerns with the appropriate authorities. And denounce Yury Adamov as a saboteur and murderer at tomorrow's daily briefing.'

'There are . . . considerations, Dr Axelrod.'

'You sound like you do not wish me to proceed, Major.'

Vasin thought of Kuznetsov's arguments. The city of cloud dwellers, the jeopardy that stalked them from the vicious careerists of the Party and the *kontora*. And he thought of Korin's words. The generals who itched to use every weapon ever made for them, who could not contemplate the end of war, forever. Axelrod was a scientist blinded by straight lines. He had spent his life negotiating the world of the concrete. His affair with Petrov was the single evidence that he had ever plunged into the tides of madness and human emotion. And he had been rejected. Then betrayed. Then bereaved. Axelrod's fury was cold, and its logic was unstoppable.

'You are right, Comrade Axelrod. You have your duty as a Soviet citizen.'

V

Nightfall brought a freezing fog, rising from the Sarovka River and creeping along the sidewalks like a ghost. Frost had settled on the trees that lined the boulevards. If clear cold is action, Vasin told himself, fog is thought. It was nearly time for his appointment with Adamov.

What he and Axelrod had found in the records clearly implicated Korin. And probably the Professor too. Even if Axelrod was wrong that his feud with Fyodor over the tamper had led to murder, Vasin knew that Adamov had another, far stronger motive for revenge against Petrov's father. And if both of them were involved, it was almost inconceivable that Masha wasn't also involved.

If it was Korin and Adamov, their fate would be out of his hands soon enough. At the very least, as soon as Axelrod made his report, the fact that Petrov had been murdered would be released from its vault and out in the world, and even Orlov himself would be powerless to lock it back in his green safe. Only Adamov could save himself now. He could somehow explain away the revelation of Korin's meddling in the files. Or he would be condemned.

The streets were nearly empty as Vasin strode toward Adamov's apartment. A couple of cars passed, but no tails were in evidence. A bad sign. The *kontora* had thrown its circle invisibly wide and almost inescapable.

In the foyer of Adamov's house an unfamiliar man sat in the concierge's lodge reading a copy of *Sports News*, or at least holding it. As he passed up the stairs, Vasin heard the soft purr as the watcher picked a telephone receiver up off its cradle.

Vasin found Adamov alone in his cavernous apartment. He wore his formal Party tunic, unbuttoned at the collar, and his decorations. He had evidently just come from a meeting with the top brass. His deeply lined face was gray with exhaustion. Without a word, Adamov led his visitor through to the dining room. Like actors in a drama, they took their old places at the table.

'Did you get your bomb built, Professor?'

Adamov's answer came as a low rumble in his throat, barely audible.

'It is done.'

The expression on the Professor's face was of an almost menacing firmness of intention. The mask that Vasin had seen slip the previous day was back in place.

'Comrade Professor, you asked me here because you have something to tell me.'

'Aren't you going to tell me that I have nothing to fear if I have nothing to hide? That used to be the most terrifying statement in the Russian language. When my interrogator said that to me, I would experience all the terrors of the abyss. And here you are, come to convict me again.'

Vasin shifted uncomfortably in the hard chair. The moment of denial came and went. He glanced toward the telephone that sat on a side table.

'Worried about your *kontora*'s little ears? Don't be. A man from the radio laboratory disconnected your pathetic bugs. He checks this place every week. I will be filing a complaint with the Committee for State Security. The things discussed around this table are not for the ears of your blundering-fool colleagues.'

Vasin absorbed the insult without comment.

'Comrade Investigator, I am not the one who harmed Fyodor Petrov. That is the truth.'

'Petrov's father's betrayal cost you your wife and daughter. They both repudiated you. No one would blame you for your revenge, Professor. A child for a child.'

'Not all things that are logical are true. With limited data, it is logical to conclude that the sun revolves around the earth. No. Arkady Petrov and I have had our own reckoning. I knew he denounced me, but to refuse to do so would have been his own condemnation. In truth, he also saved me. After my conviction, Academician Petrov officially declared that I was a vital worker with specialist skills. As a result I was transferred from the mines to a *sharashka*, a scientific workshop staffed by Gulag inmates. Some of the best work on the Soviet nuclear program was done by men like me, sitting in our padded prison uniforms. The great Sergei Korolev, who just put Major Gagarin into space? He spent the war in a *sharashka*, too. It was still the Gulag, but without Petrov I'd have died, like Korin nearly did in his hellhole in Vorkuta. So you are mistaken. I have no ancient scores to settle with Petrov, or even his son.'

'Did you ask me here to tell me that Fyodor Petrov's father saved your life?' Vasin thought of his own interview with Fyodor's father, a great man broken by grief. Indeed, he had mentioned nothing about Adamov or their shared history.

'I wanted to explain.'

'Professor, what happened to Fyodor?'

Adamov leaned back slowly into the darkness.

'You have had my answer. I did not harm him.'

'Who did?'

'He harmed himself.'

'He committed suicide? Or he brought about his own death?'

'You have a good brain, Major. And no, I am not avoiding an answer. I would say those things were synonymous.'

'No more word games, Professor. Vladimir Axelrod says that you changed the design of the tamper of RDS-220 right after Petrov's death. He believes that you have deliberately and maliciously sabotaged the bomb and wants to denounce you for anti-Soviet inclinations.'

A grim smile spread like a surgeon's incision across the Professor's face.

'Fyodor's boy. Of course.'

'He has evidence against Colonel Korin. And some compelling evidence against you, Professor.'

'That's impossible.'

'Dr Axelrod has made a computer model of the projected yield of your new design. You have deliberately cut the power of RDS-220 by at least half.'

For the first time, Vasin saw emotion on Adamov's face.

'And they were not easy to access, but the records of the calutron lab we looked at a few hours ago also proved to be quite surprising,' continued Vasin. 'The log tracking Petrov's experiments with thallium and supposedly showing that two thousand milligrams are missing? They have been doctored. Your trusted colleague Pavel Korin checked out the records the day after Petrov's death. Axelrod has found his patriotism and demands you both be punished. He is taking his information to the authorities tomorrow.'

Adamov, usually so still, jumped up and began to pace the room.

'Am I to be arrested?' His voice had become dry and bleak. 'Is this some kind of Chekist's courtesy call?'

'No, sir.' Vasin paused.

Ever since he'd arrived in Arzamas, everybody he met had told him to back off. Zaitsev, Kuznetsov, Efremov, Korin – they'd all had their various reasons for telling him to leave it alone. But what if they all had been ultimately right? The idealist in Vasin wanted the truth. The coward in him wanted salvation from the consequences of his own betrayal. Both those paths led to Korin's certain ruin, maybe Masha's too. And Adamov's. But what if Korin was right, that Vasin was just an ape in the laboratory? What if there was some great, overarching truth that he had been missing? He remembered Orlov's words. What if a crime is committed to prevent a greater crime? What if the stakes were higher than his own survival?

'No, sir. Not courtesy. But Korin told me some things. About you. About the device. The importance of your work. I need to know if there has been any . . . mistake. Before it is too late for you.'

Adamov's hands closed slowly on the high back of the chair opposite Vasin's. The Professor's hard, clever eyes roamed over him, as though searching for the answer to a question that Vasin had not posed. The last color had drained from Adamov's face, leaving only ash beneath his skin.

Adamov picked up the telephone and dialed.

'Pavel. Thank God I reached you. We have a serious problem. It's Axelrod. Yes. Fifteen minutes? Good. Vasin is here with me. Yes, him. Vasin the Chekist.'

VI

Side by side, Korin and Adamov looked like characters from a Russian folktale. Adamov was the pale wizard Koshchey the Deathless, Korin the thickset knight of old Rus. But despite the difference in their appearance – Korin shaggy but lithe with the craggy brows and faraway eyes of an explorer, Adamov desiccated and angular as a skeleton – the two men seemed to have been hewn from the same block of ancient stone. Age has made them twins, thought Vasin. The past was their common womb. Their four hands lay on the table, as lined as maps.

'You got the story, Comrade.' Korin's voice had the arrogance of a man who can no longer be bothered to lie. 'Good work.'

Korin and Adamov faced him like poker players. Vasin's cards were on the table. Now he waited to see what the old men were holding.

'I know what you're thinking,' Vasin said. 'You're going to tell me that my talents are wasted at the *kontora*. My mother does, all the time. She would have preferred me to be Yury Gagarin.'

Each of them, in his different way, smiled, though Adamov's was more a grimace of suppressed pain.

'Korin.' Adamov addressed his companion. 'We need to tell him.'

'He wouldn't listen. Stubborn little terrier.' Korin pointed an accusing finger at Vasin. 'Hand it to him. Even flew up to see me in Olenya. Got to the bottom of things fucking quickly. So what's the clever Chekist been saying to you? Confess everything and things will go easier for you? Haven't we heard that one before?'

Adamov shrugged.

'Vasin has discovered your work with the registry. His little bird Vladimir Axelrod in the Citadel believes I have sabotaged the tamper and am a traitor. Even ran a simulation of the new yield on the computer. He plans to present his case and denounce me to the authorities tomorrow.'

'That bastard. We should have acted.'

Korin stared straight at Vasin.

'We should have acted,' he repeated.

Vasin met Korin's eye and leaned forward.

'You mean, you should have had your thug smack Axelrod over the head with a wrench? And maybe me too? So that was your guy? The ape you had hanging around Axelrod? He nearly pushed me under a train. Said his bosses were "grown-ups". That would be you.'

Korin grunted and folded his arms.

'He *saved* you from falling under a train, I heard.'

'It's too late for all that now, old friend.' Adamov had turned to Korin, his voice hissing and urgent. 'But perhaps if Vasin understands what is going on, he will help us. Or else it's all over.'

The two old men exchanged a long, meaningful glance.

Adamov sighed, leaned forward into the light, and began to speak.

'Vasin, I commit the sin of science every day. I turn plowshares into swords. And yes, I mislead our masters because it is necessary to mislead them. Because they do not understand what we are dealing with. Look at the language they have invented. "Deceptive basing modes." "Baseline terminal defense." "Dense pack groupings." Our military planners make up words in order to pretend that we are in command of the forces we unleash. We used to have whole workshops full of people estimating explosive yields with slide rules. Now they are tabulated on our new computers, and we make neat graphs to report to our superiors in the Kremlin. These men understand quotas, but our numbers conceal a terrible truth. We are not in control of these forces. With every major bomb test, we please the Politburo by increasing our yields. Our bombs are bigger than the capitalist enemy's, therefore the Motherland is greater. Then, this spring, the order came down from the top of the Party. They wanted a hundred-megaton bomb. A good, round number. One hundred. Thunderous applause at the Party Plenum. Five thousand times bigger than the bomb that Oppenheimer made. These are just numbers to the apparatchiks. But it was insanity.'

The authority of Adamov's clear, measured voice was awesome. Korin and Vasin had become mere listeners and smokers.

'You must understand that the energy yielded by our discovery of atomic physics exceeds the energy yielded by that of the terrestrial, or planetary, physics of the nineteenth century as the cosmos exceeds the earth. When our grandfathers discovered steam, then dynamite, they considered themselves masters of awesome power. Now we have cracked atoms. We have forced atoms to fuse, as in the heart of the sun. Yet it is within the earth's comparatively tiny, frail ecosphere that mankind is proposing to release this newly tapped cosmic energy. So far, we have been lucky. The Americans, less so. Did Korin mention Castle Bravo?'

Vasin shook his head.

'Colonel Korin told me about Gatling. And he told me what happened at Totskoye. Forty-five thousand Soviet soldiers sent into a radiation zone. But he didn't mention any castles.'

'Seven years ago, the Americans planned to test a series of new thermonuclear devices on an island called Bikini Atoll in the South Pacific. The aim was to try out a new combination of nuclear fuels. More important, for the generals at least, was that this was the first time they were able to build a payload of deliverable size. The device weighed about ten tons.

'The Bikini tests of 'fifty-four were code-named Operation Castle. The first bomb was named Castle Bravo. There was no Castle Alpha. Our colleagues at Los Alamos National Laboratory, our enemies as we are supposed to call them, did all the calculations they could. They already had a fast electronic computer, back when we were just experimenting with our first. They estimated that the yield would be six megatons. The designers placed their cameras, their ships, their monitors, and the bunker that sheltered the detonating party accordingly. A distance of about five kilometers from ground zero.'

Adamov had a good teacher's flair for narrative. He paused now, measuring the effect of his words on Vasin.

'The device detonated perfectly. The fireball spread over seven kilometers wide in less than a single second. But according to their precise calculations, they had only expected the blast to maximize at two point five kilometers. Your spies tell us that the explosion was visible four hundred kilometers away. In less than ten minutes the mushroom cloud had reached an altitude of forty thousand meters and a width of over one hundred kilometers. The fallout was spreading at around one hundred meters per second. The firing party had to evacuate their bunker to avoid being fatally irradiated, which meant running across open ground to a helicopter and flying to a nearby battleship. They barely escaped with their lives. God knows what shape those men are in now. They say the fallout fell like gray ash, and the Pacific islanders ate it thinking it was the snow they had seen on television sets the Americans had bribed them with. Perhaps it is true. When the Yankees flew over the site a few days later, they saw that Castle Bravo had blown a crater two hundred fifty meters deep and two kilometers across. And the yield? Fifteen megatons, not five.'

'How did the Americans get their numbers so wrong?' Vasin's mind was reeling from the incredible scale of the destruction that Adamov was describing.

'No, Major. Their calculations were as accurate as our own. Their mistake was making assumptions. And their assumptions were wrong. They used lithium, the lightest of all metals, as their source of tritium, a kind of heavy hydrogen. They assumed the lithium 7 isotope they used in the second-stage bomb would be inert, that it would react harmlessly and turn into bromine. It didn't. Their new lithium-deuteride fuel produced much more tritium than they expected. The point is that despite all the previous real-world tests they had done, all the calculations they had run on their new computers, all the efforts of those Golden Brains in the New Mexico desert, Castle Bravo was three times more powerful than intended. Three times, all because of a single wrong assumption. What happens if a device is one hundred times more powerful than expected? Fifteen years after Hiroshima, and we have already achieved a five-thousandfold increase in the power of our bombs. Why not fifty thousand? Five million?'

Adamov looked at Vasin as though he were expecting an answer. Vasin had none.

'How could our calculations be one hundred times off? Or one thousand? Here we come to the most terrible secret of all. Most of the nuclear material in any device does not detonate. This has been the case since the very earliest tests. Dr Oppenheimer's first gadget, Trinity, was the world's first nuclear explosion, back in 'forty-five. They used more than six kilograms of plutonium-gallium alloy as the bomb core. Oppenheimer himself predicted it would yield less than half a kiloton of explosive force. Enrico Fermi offered to take bets among the top physicists and military present on whether the atmosphere would ignite, if it would destroy the state, or if it would incinerate the entire planet. Trinity surprised them all and generated twenty kilotons of explosive force.

'But that's not my point. Out of six kilograms of plutonium, only around one gram – one single gram – actually reacted during the explosion. The weight of a kopeck. Almost six kilograms of plutonium, puff, blown to smithereens, wasted, expelled into the atmosphere. Most of what we call nuclear fallout is merely unexploded bomb fuel. And so it has been ever since. Ninety-nine point nine percent of the plutonium and uranium we put into our devices is not detonated. The energy released by the nuclear reaction destroys the metal tamper that is built to contain it within milliseconds, and therefore destroys the conditions under which the reaction can continue. So we tried to build a tamper that contains the reaction for twice as long. Longer. What happens if not one six-thousandth of the fuel explodes, but one sixth? Can we contain the explosion long enough for *all* of it to detonate?'

'You're talking about Petrov's solid uranium tamper.' Vasin could not hide the tremor in his voice.

Adamov pursed his lips.

'Petrov designed the strongest tamper ever made. And one that was itself made of potentially fissile material. RDS-220 was meant to be a hundred megatons, seven times more powerful than Castle Bravo. At that level, the chain reaction may not stop with the hydrogen inside the device. What if it continues to the hydrogen in the air? There is a chance that this thermonuclear chain reaction could ignite the hydrogen in earth's atmosphere. Set the world on fire, Vasin, just like Fermi joked. And that, I repeat, is a danger we have been considering even without any unpredictable effects from the uranium tamper. Even without more of the fuel reacting than in previous tests.'

Korin took over the narrative, looming into the circle of light on the table as he began to speak.

'Petrov's uranium casing would become a bomb in itself. The ambitious little cunt. He wanted to ride RDS-220 all the way to the Academy to join his father. Then on to the Politburo. Wouldn't listen to our objections. "In this game you never know until you try. Old man Oppenheimer, when he detonated the first atom bomb in America back in 'forty-five, wasn't sure he wasn't going to blow up the world." That's what Petrov said. Then he laughed.'

'Professor – why did you not just overrule Petrov?'

'Petrov threatened to take it to his father and the Presidium of the Academy of Sciences.' Adamov's voice rose in indignation. 'Useless bunch of careerists. He saw the calutron readings and he knew the danger. But "the risk is acceptable if we are to show our capitalist enemies that the USSR commands the heights of nuclear supremacy". Fool. If RDS-220 sets off an uncontrollable chain reaction, the only heights he would be commanding would be a smoking pile of ash where the Kremlin used to stand. Petrov threatened to denounce me for subverting the project. Said that I was no longer suitable to lead the nation's weapons program. That I was an enemy of the people. That when he was in charge, there would be two- and three-hundred megaton devices! He said that I could choose to read about his triumphs in a retirement dacha outside Moscow or . . . or from a newspaper in a shitter in the Gulag. He said . . . he dared to say . . . that I was used to prison life, so I had nothing to fear.'

Adamov went silent. His thin hands had found each other on the table and seemed to be strangling the life out of each other, though his face remained expressionless.

'So you decided to kill Petrov.'

'I killed him.' Korin's voice was as heavy as a cudgel. 'Adamov came to me wringing his hands. Just like he is now. Petrov had to be stopped and I promised Adamov that I'd stop him.'

'And save the world?' Vasin had meant the words flippantly. But as he spoke them it occurred to him that they might literally be true. 'You invited him to tea to discuss the tamper. Maybe you even promised to see his vision through. And then you poisoned him.'

'Adamov knew nothing. I got the thallium through my own channels.'

'I saw Petrov's body. How did you handle it and not get poisoned yourself?'

'Thallium emits a radiation that is too weak to penetrate human skin. It's harmless, unless it is ingested. Two grams is just dust in a piece of paper. I slipped the cup Petrov had used into my pocket so Masha wouldn't accidentally poison herself.'

'Then you altered the laboratory records to make it look like Petrov had signed for the thallium himself.'

Korin merely raised a shaggy eyebrow in answer. Adamov's face had tightened into a grimace of pain.

'Why are you telling me this?'

Adamov and Korin exchanged a glance. The Professor, as the senior man, replied for both of them.

'We need to talk to Axelrod, tonight, and tell him the same things we've just told you. He's a man of science. He must see reason.'

'So order him to come over.'

'He won't come for me. He loved Petrov, we all knew. But he's your stool pigeon. You can make him come. Tomorrow Axelrod will be at the Party headquarters dictating a complaint to the Ministry. The next stop will be your *kontora*. It's only in our hands for a few hours more. They will halt the test and rebuild the tamper with uranium. And we cannot predict what will happen then. As for us, who knows what the Party will decide.'

'That's obvious enough. The Party will decide that you're murderers, gentlemen. The altered lab inventory, with your signature on it, Colonel, is your death sentence.'

The two old men held Vasin with their stares, waiting on his word. Vasin had heard a lot of lies in his time. But told in Adamov's measured whisper, this story sounded very true. For a delirious moment, Vasin thought, I have witnessed the beginning of the world's end. If I do nothing, the bomb will pass to other hands. And it will be uncontrollable. Armageddon will have started here, with these men, at this table, with my inaction. Images from the film of RDS-100 played in his head. The furious tidal wave of wind, the blinding light, the titanic pulse of the cosmos's energy unleashed on the fragile earth. Destroying everything. And everything he hated. Vera, turned into a pillar of ash. The *kontora* and Orlov and Zaitsev, the whole rotten edifice of lies that he served incinerated, burned clean away. And his own worthless self, of course. Annihilated. It would be a kind of answer. An end to everything, forever. Except maybe the *kontora*'s files, in their deep bunkers.

But then he thought of Nikita at school, with the windows exploding and the air turning to fire. The split second of alarm in his childish face. A bubble of nausea like an air pocket filled Vasin's lower chest. It seemed that something had gone wrong with time. The past and the future huddled in the present, their impossible weight pressing in around the three men at the table.

Adamov was convinced that he could talk Axelrod into silence, that he was a rational man. But Korin had murdered the love of the young scientist's life. Fedya, the boy with the sun in him, Axelrod's sin, his soul. Axelrod had been willing to risk everything with his desperate act of blackmail to keep Petrov by his side. Vasin knew from his old cases just how terrible the cold fury of a bereaved lover could be. What could Adamov possibly say to Axelrod that would make up for the violation they had inflicted on his life? Petrov had believed that his uranium tamper design was safe. Therefore Axelrod did. With Petrov dead, that belief might become a way of keeping faith with his murdered lover. Perhaps Axelrod had become convinced that Petrov's idea should live, now that he did not. Vasin had seen flashes of a fanatic light in Axelrod's eyes. Would Adamov's authority be able to extinguish it? Was Adamov capable of understanding love and loss in another after he had cauterized feeling so thoroughly in himself? And if Adamov failed to persuade Axelrod into silence, then, clearly, Korin would have to take action.

They were asking Vasin to lead Axelrod to his death. Somewhere in this sleeping city, Korin's sailor was slumbering on some stinking cot, ready to answer the call of a ringing telephone. If Vasin refused, or failed, to bring Axelrod to their agreed rendezvous at the registry of the Institute, Korin's man would come for him. The only way that Axelrod would live to see another day was if he spoke to Adamov, and agreed with him. According to the new logical web that Adamov and Korin had suddenly thrown over his life, the path was clear. Vasin was fetching Axelrod not to his death but perhaps to his only chance of life.

Perhaps it was not too late.

Vasin made his decision. A plan that might save Axelrod began to hatch in his mind. And Masha was the key to it. He stood.

'I know how to get Axelrod. But we do it my way.'

Korin lumbered to his feet to face Vasin, both their faces in the darkness cast by the lampshade.

'Bring him to the Institute. We meet at the registry.'

Vasin could find no obvious flaw in Korin's logic. And the Citadel, even at night, was at least a semipublic place, safer for Axelrod.

'Very well. The registry. But you must bring Masha. I insist that she be there. No arguments.'

Reluctantly, Adamov and Korin nodded.

'One final condition. No one dies tonight. No one.'

PART FOUR

I AM BECOME DEATH

We knew the world would not be the same. A few people laughed. A few people cried. Most people were silent. I remembered the line from the Hindu scripture the *Bhagavad Gita* . . . 'Now I am become death, the destroyer of worlds.'

ROBERT OPPENHEIMER, REMEMBERING TRINITY,
THE FIRST ATOMIC BOMB, JULY 16, 1945

CHAPTER TEN

SUNDAY, 29 OCTOBER 1961

ONE DAY BEFORE THE TEST

I

Vasin swallowed the chill air of the street like a draft of cold water. The lamplit faces of Korin and Adamov, hazy in the cigarette smoke-filled dining room, swam before him. The evening's fog had cleared, leaving the air damp and fresh. He looked up and saw a young moon swimming among wreaths of cloud and moist stars. The plaster caryatids of the building's facade looked down on him like old gods, trapped in cocoons of peeling paint. Vasin had walked into the building a man of truth, filled with righteous questions. He had emerged two hours later something quite different. A conspirator.

He sensed rather than saw movement, left and right. Car radio receivers were being lifted off their cradles, whispers were traveling in waves through the air around him. Like a magnet trailing iron filings, Vasin's tails were assembling into formation.

It was just past midnight. The work of the scientists was over and Axelrod didn't seem the type to celebrate at Café Kino. RDS-220 had, earlier that day, been carefully hauled up from its cellar and into the daylight. The device was now in the hands of the engineers, the pilots, and the firing crew. Somewhere, under this same moon, the railway flatcar bearing the bomb was trundling slowly northward through the snowfields of Central Russia, heading for the railhead at Murmansk. It would then be transferred to a truck for its final terrestrial journey to the Olenya air base. Korin was due to leave by plane to rejoin the precious cargo in eight hours' time. The test was scheduled for the following morning, Monday, at 11:30.

Only a moment of hesitation remained before Vasin threw his lot in with this pair of madmen. But what was the alternative? To put nine grams of lead in the back of Axelrod's head, as the likes of Zaitsev, Orlov, and the other old butchers of the *kontora* would have done, without hesitation? To run to Efremov and tell all? To have Axelrod taken into protective custody? To hand Zaitsev a triumph, standing in the witness box at Adamov and Korin's trial? To count the weeks and months before another scientist built an RDS-221, mightier and more deadly than its sabotaged predecessor? And so on until somebody finally succeeded in consuming the whole world with fire? Or maybe Adamov's version of RDS-220 could still go out of control, like Castle Bravo, and incinerate them all.

A chill breath of wind returned Vasin's thoughts to the present. He found himself shivering. How long had he been standing here, paralyzed by thought, before his large audience of invisible watchers? He shook himself into motion. But rather than heading to Builders' Street, he turned toward home. For the first time in his career, he needed his gun.

II

The apartment's windows were dark. Vasin crept up the stairs and gingerly turned his key in the lock. He could see no sign of Kuznetsov's usual discarded boots and coat in the slice of light that fell from the landing into the hallway. Vasin slipped off his shoes outside the front door and padded down the corridor in stockinged feet. The glow of the streetlamps, eerily orange when filtered through the curtains, gave him just enough light to see the outlines of the furniture. Kuznetsov's drifts of magazines and books had disappeared. There were no empty teacups, ashtrays, or dirty glasses on the coffee table in the sitting room. The sofa and two easy chairs had been aligned at precise angles. Even Kuznetsov's record collection was in wholly unaccustomed perfect order on the Czech dresser. The place had evidently been very thoroughly turned over, then equally thoroughly cleaned up, by professional hands.

Vasin remembered that his door handle had always squeaked. A new spring. He began to turn it a fraction of a millimeter at a time, pulling the door gently toward himself to release the friction of the tongue in its housing. At the very first hint of the familiar metallic squeak, he stopped and listened to the silence of the apartment. Was it possible to feel people, in the dark? Could one really sense another human being without seeing or hearing him? Because Vasin certainly sensed an unsleeping presence, somewhere in the shadows. Yet each time Vasin paused, holding his breath and listening with every part of his body, he heard nothing. He continued to turn the handle until he felt it press home against its restraining latch. He swung the door open.

Vasin's bedroom windows opened on the courtyard, and there was less outside light to illuminate the room. He made out a pile of his dirty shirts at the foot of the bed and the white detachable collar of his uniform tunic picked out on the dark bedspread. On the back of a chair by the window, he could see his leather service belt, hanging taut under the weight of his pistol in its holster. He took a step toward it but stumbled on his tall boots, hiding treacherously in the shadow of the bed. Vasin froze and listened. He took two more steps and fumbled with the fastening of the holster, slipped his hand around the grip of the heavy Makarov, and pulled it out.

The bedroom light snapped on. For a split second Vasin was blinded and paralyzed by the brightness. Recovering his senses, he spun around into a crouch, his pistol pointed at the door.

Kuznetsov stood in the open doorway. He wore his usual shabby civilian slacks and shirt, but his face had none of its usual jollity.

'Greetings, Comrade. Trust you had a pleasant evening. I've been waiting up.'

Kuznetsov's eyes were on the gun, but his voice was artificially cheerful. Vasin put his finger to his lips and motioned Kuznetsov into the corridor with his eyes. He lowered the Makarov and stuffed it into his trouser pocket. Vasin gestured his handler toward the stairwell. Kuznetsov shuffled down the corridor more in protest than obedience.

'You're a fucking wrecking ball, you know that?' Kuznetsov hissed. 'The electricians were here today, installing bugs. They took my goddamned books. My *Vozdushniye Puti*. My Mandelstam. The Lubyanka library requires them for an audit, my ass. So thanks very much, Comrade Vasin.'

'I'm sorry to hear that, old man.'

'I've never had a gun pulled on me. Got to say, didn't think you'd be the one to take my virginity on that score.'

Kuznetsov folded his arms across his chest and faced Vasin across the stairwell.

'I've listened to you a lot, Kuznetsov. Now you have to listen to me.'

'I was afraid you'd say that. Your face, when you waved that tool about. Doesn't look like a late night game of cops and robbers. Now you're going to tell me I shouldn't mention that to the duty officer when he calls after you go back out. Which is where I presume you're heading.'

'Yes. That's exactly what I want you to do. Just say I took a shit, then left on official business.'

'With your gun.'

Kuznetsov stroked his beard, grimacing doubtfully.

'This is a very long shit you're taking.'

'Please.'

'You planning to put a hole in anyone tonight with that thing?'

'I beg you, Kuznetsov. I heard everything you've told me. What I have to do tonight is not going to hurt your cloud dwellers. Please. Just keep quiet.'

Kuznetsov demonstratively put his hands to his ears and began humming.

'If you discharge that cannon of yours, I'm in for the high jump. I'm under specific instructions to keep an eye on everything, including your sidearm.'

'I won't. Thanks, friend.'

Vasin, in a gesture of intimacy that did not come naturally to him, clapped Kuznetsov on the shoulder and set off down the stairs.

'Psst.'

Vasin retraced his steps and leaned into Kuznetsov's whisper.

'You forgot to flush. And, there's no fucking clip in your pistol. You might find one in the front pocket of your holster. That's what they told me during basic training, at least.'

With a grateful glance Vasin darted back inside, pulled the chain, retrieved the cartridge clip, and slammed it home into the breech of his Makarov.

'Hope your business does not detain you long, Comrade.' Kuznetsov was back in the corridor now, speaking loudly.

'You know how it is. No rest for the wicked.' Vasin clenched his fist and raised it in solidarity, a gesture he had seen in a newsreel of Cuban revolutionaries, his voice self-consciously loud and cheerful. 'They shall not pass, Comrade! *No pasarán, compadre!*'

III

Vasin turned off Gogol Boulevard onto Builders' Street. The last tram was long gone. Most of the suburb's residents, too, seemed to be parked for the night in their beds, though light still flickered in a few windows like gray bonfires. Tele-Vision machines, Vasin concluded, though he had never been in a home that boasted one, not even Adamov's. He walked on the edge of the sidewalk, in full view of the *kontora* Volgas that ritually passed every two minutes. He could guess their orders. Straddle him, but don't touch him. And don't lose him.

To his left were rows of dark trees, and a black mass of undergrowth. The streetlamps hung over him like yellow moons caged in broad steel brackets. Vasin glanced nervously from time to time into the thickets. Was Korin crazy enough to send Sailor and other cutthroats to murder both him and Axelrod under the noses of a full *kontora* surveillance team? There had been a chilling calm in Korin's confession to the killing of Petrov, the practical callousness of a military commander. Korin needed Vasin for as long as it took to fetch Axelrod. But afterward?

He hesitated in front of Axelrod's archway. In the courtyard, a lynch mob of deformed apparitions appeared to have gathered in formation, ready to jump him. Most menacing was Karandash the clown, who cast the moon-shadow of a gorilla, backed up by the big-eared bear Cheburashka. Axelrod's windows were dark. Vasin found a pay phone, positioned, just like in Moscow buildings, in the shelter of the main archway of the building, and dialed the scientist's home number. Engaged. He waited two minutes and dialed again. Phone off the hook, most likely. Vasin crept through the shadowed part of the courtyard to Axelrod's entranceway. The staircase was silent and deserted. Vasin waited three minutes in the first-floor-landing window, watching for signs of life in the courtyard, but saw none. He adjusted the pistol, uncomfortably wedged down the back of his trousers, and went up.

Axelrod's flat had a mechanical bell operated by twisting a knob in its center. It tinkled feebly in the silence. Eventually Vasin heard stumbling footsteps inside the apartment.

'Who is it?'

'Vasin. We need to talk.'

A silence.

'It's one in the morning.'

'Let me in and I'll tell you everything.'

'Are you alone? Was that thug lurking outside?'

'I'm alone.'

Another long pause. Vasin heard the grating of metal on metal, then the click of the lock turning. Axelrod opened the door a crack. But his pale face peered out through a steel door restrictor that prevented it from opening further.

Vasin had installed exactly the same restrictor at his mother's apartment. It was a solid fucker. By the time he kicked it in, the local cops would be swarming.

'It's Maria Adamova. She wants to see you.'

'*Now?* Why on earth?'

The suspicion in Axelrod's voice had given way to alarm.

'I think you know. We both know. She wants to talk to you about the information she has about you and Petrov.'

Axelrod's eye disappeared from the door crack. Vasin felt the weight of the scientist's body slump against the door.

'I don't care. She can't save her husband and her fancy life.'

Axelrod's voice came, muffled, out of the darkness of his own hallway.

'She wants to protect you. That unpleasantness doesn't have to hang over your life. Your career.'

'Now she wants to blackmail me. Those photographs for her precious Adamov's freedom. She thinks I care so much about myself that I will forget what that animal did to Fedya?'

The logic of a man in love. Vasin could think of no answer to it.

'Honestly, Axelrod? I don't know what Masha wants.' Vasin weighed his words carefully. What do you tell a jilted lover? 'Perhaps she loved Fedya, too, and she doesn't want his bright memory besmirched.'

A long pause.

'What aren't you telling me?'

'It was not Adamov who killed Fedya.'

'How? Vasin, you're lying.'

'Just come with me to see Masha, and you will learn everything. She has all your answers.'

'Bullshit.'

'Masha is desperate. She will not destroy the evidence unless she speaks to you.'

'Why has she confided in you?'

'Because she thought *you* poisoned Petrov. But today I told her about the lab reports that Korin falsified. She realizes she has made a grave mistake. But she needs to see you. Now. Before the morning.' A final flash of inspiration sparked in Vasin's tired brain. 'She fears that you are too blinded by hatred for her husband. To listen to the real story. If you do not come, you will never know. And you will send the wrong man to perdition.'

Nothing from behind the door. For a desperate moment, Vasin feared that Axelrod had tiptoed away from the door and was now dialing the police. But then he heard a sigh from inside the apartment and felt Axelrod lean into the door once again.

'She swears to keep quiet about me and Fedya?'

'I can't speak for her. But I am sure that's what she wants to speak to you about.'

'And you? You will also keep silence?'

'I give you my word.'

'I hardly know you, Vasin. What's your word to me?'

'Listen. Axelrod. I know more than you think. I know how important Adamov is to the RDS program. I know his value to the defense of the Motherland. How important his genius is to all of us. What you propose to report tomorrow is only part of the truth.'

The silence stretched. Vasin imagined he heard his own heart thumping. With an effort of will, he prevented himself from thinking of what he would have to do if Axelrod refused.

The weight of the scientist's body lifted from the door. Vasin heard the restrictor grind home, then free. The door opened again, revealing Axelrod in gray pajamas and dressing gown. He looked older, his eyes ringed with exhaustion.

'God knows why I'm trusting you, Vasin.'

'Because I'm on the side of the angels, Axelrod. Always have been.'

IV

They trudged along the deserted boulevards in silence, broken only by the regular soft crunch of cruising *kontora* Volgas. Axelrod was clearly nervous. Was he afraid of confronting his lover's mistress? Masha said they had not seen each other since the evening Axelrod had run from Petrov's apartment. Or was he ashamed of suffering the indignity of having to bargain with Masha, who held Fyodor's reputation and Axelrod's future in her hands?

The Citadel was quiet, quieter than Vasin had ever seen it. In the final days of RDS-220's gestation, every corner of the place had hummed with activity. Now he and Axelrod found the building nearly silent. The building's baby had been birthed. For the hundreds of men and women who had worked so intently on its creation, there was nothing left to do but wait. It was the turn of others to see their precious device on its last journey into the skies over Novaya Zemlya.

Once through the turnstiles in the entrance hall, deserted apart from a single sleepy sentry, Axelrod turned instinctively to the lifts that would take them to Adamov's office. Vasin stopped him.

'She is waiting for you at the registry, Doctor.'

Axelrod blinked quickly, and looked imploringly at Vasin, as though begging him to reveal whether he was really some enemy who worked under the cover of friendship, trust, and pity. Finding no answer, Axelrod looked past him into some troubled private territory of his own. He hesitated for a final moment, then turned quickly and began to descend the stairs.

Axelrod was still well ahead of him when they turned in to the corridor that led past the calutron hall and the laboratory that housed Dr Mueller's barometric chamber and went on to the registry. Masha stood halfway down the long passageway. Her feet were planted apart, her hands deep in her pockets, her chin sunk into her chest. She looked up as they came into view.

'Hello, Vladimir. You came.'

'Maria Vladimirovna.'

Their voices sounded unnaturally loud in the echoing corridor. She straightened, arrogant and defensive at the same time.

'Let's go somewhere *they* can't hear us.'

Masha flung a contemptuous glance down the corridor toward Vasin. Axelrod also turned, mirroring her look. In the moment of mutual dislike of the *kontora* man who stood before them, Masha put her hand on Axelrod's shoulder and forearm.

'Come.'

She led Axelrod past the double doors of the calutron hall and on down the corridor, speaking to him in a low, confiding voice that Vasin could not hear. She steered him toward another pair of doors further down the passage, increasing her pace as she went. There was something angular and not quite reconciled about her movements. She's moving too fast, thought Vasin. He had fallen back a respectful distance, but now he began to lengthen his stride. Masha opened a door and herded Axelrod inside. As he passed before her into the darkened laboratory, she turned and fixed Vasin with a fierce stare of warning.

He broke into a sprint.

'Axelrod! Stop!'

Pushing Masha aside, Vasin burst through the door. An axe came flying within millimeters of Vasin's face and connected squarely with the nape of Axelrod's neck, a few paces in front of him. The blow knocked the skinny scientist into a stack of files and buried him in an avalanche of sliding paper. As Vasin lurched back to avoid the blow, he recognized Korin's broad back and shoulders carried through half a turn by the swing of the axe. Korin was a powerfully built man and recovered quickly. Almost without pausing to size up his next target, he raised the weapon again and swung it with all his force at Vasin's head. Vasin ducked, by instinct, and the heavy steel sang past his ear.

'Korin! You lying bastard.' Vasin glanced around him. He sensed, rather than saw, a great, dimly lit space, as large as the calutron lab. Behind him, blocking his escape route, was Masha. She had turned to bolt the doors to the lab shut behind them.

Adamov's voice, coming from somewhere in the gloom, was reedy with alarm.

'Pavel! Have you gone mad?'

Vasin's hand went to his waistband and fumbled with the unfamiliar Makarov. He turned to face his attacker.

In front of him, Korin went down into a crouch, the heavy fire axe in his left hand and the fingers of his right poised to gouge the eyes. Korin had a feral look that Vasin had seen in some criminals, the look of a man ready to do more hurt than he needed to. And Vasin had seen that fighting pose before. The hovering crouch of the *urka*, the convict, before a knife fight. Too late, Vasin remembered the next move. A scything kick that knocked his legs from under him and sent the ceiling lights spinning away with a sickening suddenness. As he fell he saw Adamov's face, pale, aghast, flash through his field of vision. The back of Vasin's skull connected with the floor, and his head exploded in stars.

V

Vasin came to in a dark world that rang with pain. He lay on a cold concrete floor, alone. His hands had been tied, quite expertly, behind his back with a rough strip of cloth that chafed his wrists. White light burst through his head when he tried to lift his face from the floor, and he could taste blood in his mouth. In his line of vision he saw the bases of steel filing cabinets, the legs of stacked office chairs, and a scattering of small black balls the size of peas. He seemed to be in some kind of side office that was divided from the main laboratory by a row of windows. One leg was bent painfully under him. Mercifully, his legs were not tied, and he was able to roll from his side onto his back. His whole body seemed to be trembling at low frequency, a hideous vibration that grew slowly and made his skull ring with pain. But when, with an effort, he lifted his head from the floor, Vasin realized that the rumble came not from inside his body but from the outside. It was the whine of a large machine, spooling up into motion.

Vasin managed to scrabble along the floor a little way. He felt something squashing under his shoulder blades, releasing the unmistakable smell of animal shit. Vasin recognized the sour, farmyard smell from the day he'd visited Axelrod and his calutron. This must be the laboratory of the little German doctor they'd brought from the concentration camps. Vasin remembered his nervous greeting as he followed his wagonload of crushed goats down the corridor.

Vasin managed to reach a wall of filing cabinets and work his way into a sitting position. The rumbling had grown louder and was now joined by the audible whine of an accelerating flywheel. He rolled over onto all fours and got to his feet, the back of his head a mass of fiery pain. The only light in the vast space beyond the glass-walled office came from a series of lamps that illuminated a kind of raised console bordered by steel-boxed, dial-studded controls. Three figures were huddled under the lights, talking animatedly in hushed voices. Unmistakably, Masha's blond bob, Adamov's bald head, and Korin's shaggy gray locks. He could hear nothing of what they were saying.

Vasin saw the shadowy outlines of an array of machines in the hall. There was a vast steel ball as large as a tramcar that reminded him of a deep-sea diving bell in one of Nikita's science books. Behind it stood a pair of gigantic steel arms, like bowed-down oil derricks. The increasing whine came from somewhere in the darkness beyond.

His head still ringing, Vasin made his way unsteadily over to a table that stood before the window. Working blind, he began to rub the knots on his aching wrists back and forth against the desk's corner. His bindings only got tighter. Cursing, he looked back to the conspirators around the console, his face illuminated in the light for a brief moment. Masha glanced over in his direction at the same second.

Vasin ducked back down into the shadow. But the voices in the hall had stopped. He heard quick footsteps approaching across the echoing hall. A key rattled in the lock, and the office door opened. Masha appeared, backlit in the doorframe. She was carrying Vasin's Makarov.

'You can stand up, Sasha. I see you.'

Vasin straightened up, reeled giddily, found his balance by leaning on the edge of the table.

'Everything's fine,' Masha called back to Adamov and Korin.

She backed against the open door and placed one foot flat against the wood. She looked girlish, except for the gun she held loosely by her side.

'How's your head? Korin didn't mean to . . .'

'Hurt me? Think he did actually.'

'I mean – you know.'

Masha puffed out her cheeks. She raised her pistol hand to scoop back a lock of stray hair behind her right ear.

'I'm sorry you got hurt. Really. It wasn't meant to be like this.'

'I told your husband that I would not bring him here like a lamb to the slaughter. He gave his word.'

'Axelrod's just one man. This is about more than that.'

'Just tell me that Axelrod is alive.'

'Yes. Axelrod is still with us.' Masha's voice had become suddenly hard.

Their eyes met. Vasin felt suddenly swept by a regret for what could have been, the refuge Masha could have given him. So she really was one of them. The knowledge of her deception ached as much as his throbbing head. Masha's stare was defiant, as if to appeal for his support in some argument she was conducting within herself.

'None of this was my idea, Vasin.'

'Not your idea, to use me? For information?'

'You came into my life. On the roof of the Kino. You saved me.'

'I can see you're grateful. What do you do to people who piss you off? Actually, I know. You stick a knife in their ribs. If that was even true.'

Masha's face pivoted, pinched with anger, toward the light for a moment before she composed herself and leaned forward.

'Poor innocent Vasin.' Her voice was an angry whisper. 'When will you stop being a child? Everything I told you was true.'

She wiped her face with her sleeve, the dull gleam of the Makarov's butt catching the light.

'I did like you, Vasin. I do. I'm not that good a liar. There. Believe me if you like.'

'But you used me. To find out what I knew.'

'Yes. But only because it turned out that you knew things.'

'You're going to tell me Adamov put you up to it.'

'Some women are actually capable of rational thought independent of their husbands. Might be news to you. I knew that you came here to find out who poisoned Fedya. It was dangerous. To Adamov. To the project.'

'You knew who killed Petrov all along.'

Masha looked away.

'You sat at the table while Petrov drank poison. You must have been within a meter of him. Probably poured his tea yourself. God knows, you had the motive. Petrov betrayed you. You wanted him dead.'

Masha closed her eyes and leaned her head back against the door. She gulped, her long neck reflecting the light.

'Yes. I did pour the tea. And yes. I did know what Korin was planning to do. I knew it would be the last time I would see Fedya. It wasn't easy. For what it's worth. But I didn't want him dead. Not for myself.'

'Sounds like you were eager enough to help. Was it Adamov who told you about the plan? Korin?'

'Adamov didn't know. I listened to their conversations around the table. I knew what Petrov was to them, how dangerous his ideas were. So when Korin privately told me that something radical had to be done, I agreed. The method was Korin's idea. He didn't want to put me in danger by keeping me in the dark about what he was going to do. As for the rest . . . as for you . . . we just made the calculations. Just like Adamov taught me to do.'

'And you calculated that you needed to keep me close.'

Masha puffed out air by way of answer. Her eyes avoided Vasin's.

'Was I right that you were in love, Sasha? A little bit?'

'Enough to spill secrets. That's true.'

'Ah.'

'Masha. Listen to me. Your husband promised me that he would persuade Axelrod. That's why I agreed to go get him. I didn't agree to bring him to his death.'

'No. No, you wouldn't.'

'But you knew all along?'

'You think I know everything, don't you? I thought that Korin would let Adamov at least try to talk to him. But Korin is a man of action. He believes destiny guides his hand.'

Masha stared at him, then with her habitual abruptness of movement stood up straight.

'One last thing. In the bathroom of your apartment. When you kissed me . . .'

Behind his back, Vasin continued working the knot with quiet determination against the corner of the desk. He could finally feel it getting looser. Just a couple of minutes more.

'Guess you'll never know, will you, Sasha?'

She sniffed violently, and once more rubbed her face with her gun hand.

'Careful with that thing. It's loaded.'

'I know it's fucking loaded. Patronizing asshole. I was top in marksmanship at my institute. Korin used to warn Adamov I'd shoot the cap off his head if he got me angry. I can handle myself.'

The knot finally came free. The cloth that had bound him so painfully had been, he realized as it unraveled, his own woolen tie. Vasin clenched and unclenched his hands to get the blood flowing again.

'If only I had a cap. But I lost it. Off the roof of a cinema.'

Masha grinned, despite herself, with genuine warmth. Vasin smiled back, willing this moment of simple complicity never to end.

Three seconds. Four. His left hand shot forward, seizing her right wrist. With his other hand Vasin cupped the back of her head and forced it forward, twisting her into a sudden headlock. A movement he'd routinely flunked at training school, executed with a

283

wholly unexpected perfection. Her hand, the finger on the trigger, was now closed in his. Muffled in the side of his coat, Masha tried to scream, but her shouts were drowned by the rising din of the engine.

Vasin squeezed the pistol out of her sweating grip. Defeated, she seemed to relax, her hand dropping around his waist in a bizarre parody of an embrace.

Vasin shifted his weight around Masha so that he could step out of the doorway when he released her. He was ready for her to spring at him like a banshee when he loosed his grip on her head. Instead she only slumped backward, rubbing her neck.

'Shit.'

'Did I hurt you?'

Vasin was outside the doorway now, keeping her covered with his pistol.

'Vasin, please. None of this was Adamov's idea. It was all Korin. And me.'

'Put your hands on your head.'

'Or you'll shoot me?'

'If I have to.'

She tossed her head contemptuously.

'No. You won't.'

Both of them knew she was right. Nonetheless she obeyed him, placing each hand on her head with exaggerated formality, like a teacher demonstrating the move to a class.

'Now walk in front of me. Slowly. And keep quiet.'

Masha started briskly across the machine hall, though she kept her hands in place.

In the dim light that came from the control console, Adamov and Korin looked gaunt and pale as figures from a church mural. Their debate had reached some kind of resolution and they stood together in silence, contemplating the controls before them.

'Sorry, Korin,' Masha called out with forced brightness. 'He got loose. I think he's got some questions for you.'

The two men looked up in alarm as Masha and Vasin approached.

Coming closer, Vasin saw that Adamov's already gaunt face had become a pale mask of shock.

'What have you done with Axelrod?'

'First put the gun down,' Korin said. 'Then we talk.'

'You must be joking.'

'Lower it, at least. Masha, come here.'

Without looking back to see if Vasin had complied, Masha walked over to join her husband. She did not touch Adamov, or look at him, but merely stood very close. Korin leaned over the controls and turned a dial.

'Stop, Colonel. Whatever you're doing. Stop it.'

But Vasin could not summon enough power into his voice to command the likes of Korin. The dull noise rose to a deafening thrum as the electric engine revved up to full power.

'Where is Axelrod?'

Korin pulled himself to full height to face Vasin.

'I'm unarmed. See?'

Korin opened the battered sheepskin flier's jacket that he was wearing over his uniform and slipped it off his shoulders. There was no holster on his belt. He raised his hands and twisted them back and forth, like a magician, demonstrating that they were empty.

'God damn it. Answer the question. What have you done with Axelrod? And switch that noise off. Or I'll put a round right into those controls.'

'Vasin, be calm.' Adamov's voice was rasping and dry, but its volume cut through the din. 'Your gun will not save Axelrod now. But now there is nothing to be done for him. It is decided.'

'It is decided? You are going to kill him.'

'He was a promising young man. Believe me, if there had been a way . . .'

'Is that what you meant by persuasion, Professor? A fire axe to the back of the head.'

The two old men exchanged a glance. Adamov's was cold and judgmental. Korin looked down, either chastened or exasperated.

'I did not authorize the Colonel to raise his hand against

Axelrod,' Adamov said. 'But he did. And now there are no longer any other possible courses open to us.'

'You think you can batter a prominent scientist to death and there won't be any questions? You think another murder will help keep your secret safe? And what about me? You plan to make me disappear too?'

'Nobody will batter anybody to death, Major. And do not forget what we are doing here. You seek the truth; today we do the work of the Lord.' Korin spoke firmly, with the emphatic authority of a commander under fire giving orders to a bickering platoon.

'Trying to persuade Axelrod to keep quiet would have been a waste of time,' Korin continued. 'And time is exactly what we do not have. Tonight Dr Vladimir Axelrod will commit suicide. And you, Vasin, the last person to be seen with him, up there at the turnstiles, you will report on his desperate state of mind as you left him here.'

Korin's face, up-lit by the dials of the console, had an air of demoniacal conviction.

'Why on earth would I testify to such a thing?'

Now Masha answered.

'Because you believe, Vasin. You know why Fedya had to die. It's all a calculation. That's why you are here. You know why nobody can ever find out. You became one of us the moment you went to fetch Axelrod.'

'I didn't bring Axelrod here so that Korin could murder him.'

Masha's gaze was cool and level.

'Adamov and I didn't come here to kill him either. And yet here we are. When you live with wolves, you howl like a wolf.'

As Masha spoke, Korin retreated slowly from the console. The bright light released him into the shadows, and his figure moved closer to Adamov.

'Stop moving, Korin. Shut down that bloody din.'

The old man said nothing, but stood stock-still in the half darkness. Vasin saw the whites of all their eyes suddenly focus on something over his shoulder. Keeping his gun trained on Korin, Adamov, and Masha, Vasin glanced behind him.

The four-meter-high steel sphere dominated the hall. Vasin remembered what Axelrod had said: 'In this cellar, we separate streams of atoms. In that cellar, Mueller explodes farm animals.' In the front of the chamber was a circular pressure door, like on a submarine. And in the center of the door was a single window. The interior of the giant ball was illuminated in ghostly red light. And in the light a face had appeared.

Vladimir Axelrod.

Blood was running down one of his temples, and he stared through the thick glass as though from the other side of life. He began to batter the steel, but no sound escaped the hermetically sealed ball. On either side of the apparatus a pair of enormous pistons, each the size of a car, were slowly rising into the air.

'Korin! What the hell are you thinking? Adamov! Masha! This is madness.'

'Not madness, Vasin. You know this has to be done.'

'It's murder. Stop.'

Korin and Adamov did not move, but continued to stare as though mesmerized at the pleading face. Masha had put her hands over her eyes and turned away.

Vasin ran over to the steel sphere, stuffing his pistol into his pocket as he sprinted. Nobody made any move to stop him. The circular wheel sealing the door spun with surprising ease. Vasin rolled the bolts all the way open and pushed at the hatch. It did not budge, but a tiny squealing hiss from inside the door mechanism told him that air was escaping the infernal machine. Some kind of pressure differential was sealing the door shut from inside. He straightened from his efforts to open the hatch to face the window. The terror that Vasin saw in Axelrod's eyes was naked and desperate. Again he shouted and banged with the palms of his hands on the glass, transported by panic. But nothing could be heard. Vasin noticed that the young man was now bleeding from both ears.

He pulled out the pistol once more and ran back to the control console, banging past trolleys and stumbling over rubber pipes that snaked across the floor in the darkness.

'How do you stop it?' Vasin waved his Makarov wildly at Korin. Korin folded his arms tightly across his chest.

'Leave it, boy. Adamov, don't watch. Masha! Turn him away.'

Maria took her husband by the arm, and he allowed himself to be steered away from the sight of the desperate struggle in the tiny window.

'Stop it, Korin!'

'I can't. The poor doctor already set the mechanism in motion. He left the door to the chamber open while he switched on the machine. Then got in himself. He knew that the plug door would seal itself as soon as the pumps began working. No way to change his mind, once he was in. A nasty way to go. But at least the choice was out of his hands. Brave, if you think about it. He knew he would never summon the will to pull a trigger or jump off a building. This way, it would be certain. The boy knew himself. Knew his own weakness. That's what you're going to conclude, Major.'

'He's alive.'

'Not for long.'

'You're lying. There's a way to stop it.'

Vasin stepped up to the console. An incomprehensible array of dials, levers, and gauges spread before him. He began desperately toggling every lever within his reach.

Korin stepped into his path, blocking his progress down the control panel. The old man stood before him, craggy and immovable as an ancient tree.

'Steady.'

'It's murder, Korin. It's my duty to stop you.'

'It is sacrifice. "God called to Abraham. He said, 'Take your son, your only son, and sacrifice him there as a burnt offering.'"'

'The fucking *Bible*? Have you taken leave of your senses, Korin? Adamov, you must stop this.'

Framed in shadow, the Professor found his voice.

'One more life. To end war, Vasin.'

The pistons rose to their zenith, and a loud klaxon sounded. Vasin, alarmed, looked for the source of the sound. Korin lunged forward and grabbed Vasin's wrist, slamming it down on the hard edge of the console. The pistol skittered across the floor, and Vasin scrambled to recover it in the black pool of shadow under the table. Korin remained where he was, guarding the console. Vasin found the cold metal of the Makarov and trained it on Korin.

'Switch it off. I'm warning you.'

Vasin snapped a round into the chamber and flicked off the safety catch. The bearded colonel loomed over the control panel, covering it with his body. Ignoring Vasin, he began to chant in the powerful, singsong voice of an Orthodox priest reading the lesson.

'For whoever wishes to save his life will lose it,' Korin chanted in his deep baritone. 'But whoever loses his life for My sake, he is the one who will save it. Glory to you, Lord. Glory to you. Glory to you.'

The klaxon continued, deafening as an air-raid warning. Red lights went on all over the control panel. At the same moment Vasin pulled the trigger three times in rapid succession. One round hit Korin in the shoulder. The other two caught him square in the chest as he fell.

With a titanic thud the pistons released, slamming thousands of atmospheres of pressure into the steel chamber. Vasin turned to see Axelrod's body burst like a popped balloon, his head collapsing and disappearing from sight.

Masha screamed and ran to catch Korin's heavy body as it fell. As the cordite smoke cleared, the only sound was the compressor's engines spinning slowly to a standstill, and Masha's thin keening.

VI

Vasin's brain scrabbled for comprehension like a dog slipping on ice, but he could find no purchase. The scene in front of him was unreal, as though he were watching his own life unfold frame by frame in one of Adamov's slow-motion films. The gun in his hand was impossibly gravity-laden. His hand fell to his side from the weight of it. Masha had gathered Korin's body into her thin arms. His lifeblood was spreading in a monstrous black stain across his tunic. Korin coughed, spitting blood messily on Masha's face, shuddered mightily, and then slumped back. His handsome gray-bearded head lolled, the mouth falling open. Adamov remained utterly still, his face blank with shock.

Masha pressed her face into Korin's, and she tried to lift his lifeless head from the floor in both hands. A stream of soft, unintelligible words came from her mouth, addressed privately to the dead man. Masha's muttering stopped, as though she was waiting for an answer. She shook Korin's head, first gently and then with increasing anger, like a child trying to shake a broken mechanical toy into life.

Once, Vasin had been a man afraid of chaos. Now chaos had embraced him. He had killed a man. Vasin turned the words over in his mind, unable to fit them into any meaning or feeling that he could comprehend. Korin, so deep-rooted and indestructible, so thickset and permanent, lay lifeless before them. Korin, the man who saw into the heart of things, the man who always had the answers, was gone. Vasin felt a stab of absurd anger. Look what you made me do, you stubborn old fucker. Are you happy now?

Vasin knew, with the violence of a physical blow, that his own life was now over. In a mad moment he had abandoned all reason to follow a pair of old lunatics into their insane plan. And now it was he who was left holding the smoking gun. The bomb, the end of the world, the uranium tamper, all the awesome, terrifying things that Adamov and Korin had told him seemed just a fantastical web of shadows, banished from his consciousness by the gunshots like shades before the light. He felt a stinging sensation between his thumb and forefinger from the pistol's kick. He saw the dead man before him, the young woman cradling the corpse like a scene from an old Italian painting in the Pushkin Museum. He tried to wrench his mind beyond the scene before his eyes. But Vasin's brain refused to obey him.

Crouching by the body, Masha released her burden. Korin's head hit the floor with a hollow thud. The sound of skull hitting concrete, so human and so physical, broke Vasin's paralysis. The outside world that surrounded them came crowding suddenly into his mind. He turned his head, listening for footsteps in the corridor, but heard nothing except the ringing of the gunshots in his ears. The engines had stopped spinning, and the pistons were subsiding with a soft, oily sigh. From the barometric chamber came a hiss of escaping air.

Vasin crossed the hall to the steel sphere. He pushed on the hatch, hard, until it finally yielded with a rubbery slurp. A single caged lamp on the inside turned from red to green as the pressure equalized, illuminating the contents of the sphere in a ghastly, theatrical light.

Axelrod lay sprawled in a pool of blood, his body akimbo like a loose sack of laundry. He looked as though he had been stamped on by a furious titan. Axelrod's head had partially caved in, and his chest was hollow. His life had been extinguished so violently that lines of black blood had spattered across the chamber's walls, trickling downward like flung paint. Vasin turned away and walked slowly back to the console, wading through the thick darkness as though it were water running against him.

Masha had straightened up, though she still crouched on her haunches. Her breath came in shuddering sobs, and her face glistened wet in the yellow lights of the console. She swayed a little, balling her fists into her eyes for a long moment. Then she pulled herself together and stood. Adamov, his own spell of immobility suddenly broken by his wife's movement, stepped toward Masha. He gathered her into his chest in a gesture that was so simple and intimate that the Professor suddenly seemed to have sloughed off his stern former self and become a vulnerable old man.

On the floor between them, Korin's body jerked in a violent spasm that lifted his hands in a momentary, shocking convulsion before they fell back down with a lifeless slap. All three started in alarm. Adamov and Masha broke their embrace as they all stared at the corpse, waiting for more movement. The Lazarus reflex, the final paroxysm of a dying body. Vasin had heard of it but never seen it for himself. Korin's skin had turned papery and deathly pale.

Masha was the first to break the silence. Her voice was parched. 'Korin sacrificed himself. He's the lamb. The sacrificial lamb.'

Vasin looked at Masha dumbly.

'Don't you see? He offered himself. He was a believer in God. Don't look so shocked.'

'Sacrificed himself, for what?'

'For us. For you. You heard what he said. "Take your son, your only son, and sacrifice him." '

Vasin shook his head, but no coherent thoughts came into his brain.

'What are you saying?'

'We have to do what he said. He sacrificed his life, now we have to sacrifice his name.'

'How?'

Masha cleared her throat. Her voice became steadier as she spoke.

'We tell the truth. Korin poisoned Fedya. Korin forged the lab reports to make Petrov's death look like suicide. And it was Korin who killed Axelrod. All that is true. Korin took his guilt upon himself. We can explain everything now.'

Abruptly, Adamov moved across the dais and sat down heavily on one of the operator's chairs. It was as if he had been folded up by some large invisible hand.

'Masha. We can explain everything, except *why*. Why did Korin kill Petrov? Or Axelrod?' Adamov reached into his tunic pocket and produced a *papiros* cigarette, lighting it with a slightly shaking hand. He spoke across the semidarkness to his wife as though they were alone. 'Child – how can we possibly explain Korin's motive? Without revealing the truth about why we made the changes to the device? And how do we explain how poor Korin ended up dead on the floor, shot through the heart with a *kontora* bullet? No, it ends here. Korin's whole scheme? A desperate gamble. He thought he could protect the world from my bomb. To protect me. But he lost his gamble. We spilled the blood of young men in vain. There is no story to explain this.' Adamov gestured to the body on the floor. His voice had become a bleak whisper. 'No. My love. We are lost. I

292

am lost, at least. If Vasin agrees to protect you, Masha, you can still run. Save yourself. Tell them that you knew nothing . . .'

'Wait.'

Clarity came to Vasin like the shivering flush that follows the breaking of a fever. A recent memory had come looping vividly into his head with the force of a revelation. A cold night, creeping along the outside of Korin's barrack. A glimpse of Masha through the papered-over window, huddled by a kitchen cupboard. The light of an electronic apparatus illuminating her face. And the thin, metallic voice carrying across the radio waves from distant capitalist lands, 'This is the Voice of America . . .'

Finally Vasin's reason had begun to make connections. His investigator's mind began to fit the pieces together as he spoke.

'Korin was a spy.'

Adamov exhaled smoke contemptuously.

'Have you lost your mind, Chekist?'

'We have evidence. Material evidence. Korin had a hidden private shortwave radio set up in his hut. He listened to transmissions from America. "This is the Voice of America." Didn't he, Masha?'

After a pause Maria nodded slowly.

'Masha? Have you gone mad?' Adamov flung his cigarette away in disgust. 'Whatever this man has promised you, it's all lies. Don't repeat his fantasies. I know what these people will do.'

'No, husband. Vasin is right. Korin did have a radio. Here in Arzamas. He put it together himself. He used to listen to American programs. Sometimes he caught a Christian radio station run by some Russian émigrés from somewhere in Canada. Lots of different voices. All clamoring for his soul. Voice of America, Radio Liberty, Voice of Israel. Maybe he heard the voice of God there too. He taught me how to use it. I would listen to news programs. Sometimes. When he was away. But mostly because it was like listening to him.'

For the first time Vasin saw Adamov at a loss. The Professor rubbed a hand across his stubbly scalp.

'That fool,' the Professor said, almost to himself. 'Saints and angels. Bloody fool.'

Vasin could see it now, the loose threads of the story tightening into stitches.

'Korin told me he worked with Americans, during the war. There was a pilot he was friendly with. Dan . . . Bilewsky. Bilewsky is the man who recruited him. Back in 'forty-two. He nursed his hatred for Soviet power through his years in the Gulag. And after he was pardoned for his crimes against the Party, he insinuated himself into the Motherland's weapons program in order to betray it.'

A ragged sigh of disgust came from Adamov.

'Chekist, you know your job too well.'

'No, Comrade Professor. I know their minds well. It's not about finding the truth, it's about telling a story the people in power will believe. You will be questioned. You will say that you guessed at Korin's secret religious sympathies. You will say that he often expressed anti-Soviet attitudes.'

'You want me to denounce him.'

'Yes. You will denounce a dead man. As he would have wanted you to. And the bomb, your version of the bomb, will drop on Monday morning without setting the whole damn world on fire.'

Adamov had recovered some of his icy spirit.

'And how does Korin's holy radio explain . . . what we have here?'

'Korin knew Petrov liked foreign films, foreign literature. Decided that he would be susceptible to treachery. Korin tried to recruit Petrov. But he went too far. Every attempted recruitment is a calculated risk. Korin had to expose himself, reveal what he was. And when Petrov refused, he had to be dealt with.'

'So you will say that Korin the traitor murdered his young, brilliant colleague just to protect his own hide?'

'Exactly. Then he forged the record to make it look like Petrov committed suicide.'

'Quite the snake, this Korin of yours. And Axelrod?'

'Axelrod knew Petrov well. Very well.' Vasin shot a glance at Masha. 'They were lovers, in fact. Axelrod suspected that his friend's death was not suicide and came to me with his suspicions.

But it was only when he and I checked the laboratory records together that we found Korin's name on them.'

'And where does Sherlock Holmes come into this? I mean you, Major.'

Vasin ignored Adamov's sarcasm.

'Korin was present at the dinner where Petrov was poisoned. I heard him listening to American radio. When I interviewed him in Olenya and here in Arzamas, he was defensive and told me many subversive stories against Soviet power.'

'Richard Jordan Gatling?' Masha piped up. 'Marshal Zhukov's nuclear test on our troops?'

'All that. Yes. I tell the *kontora* I had strong reasons to believe that Korin was a dangerous element. And then, when Axelrod and I found out about the forged records, I decided to bring this information to you, Professor. Privately. You were shocked. Korin was your old friend and colleague. You wished to hear this story from the mouth of his accuser, Axelrod. So you asked me to bring Axelrod here, to the Institute, tonight. And then you made a fatal mistake.'

'I told Korin?'

'Yes. You called Korin. No point in denying it. The *kontora* would have listened in to the call; it went through the central exchange. You could not credit what I told you. Your impulse was one of loyalty to an old comrade. You regret it now, of course. But you could not believe in the Colonel's treachery. And Korin was such a very good liar, to survive all these years in the heart of our most secret city. Such a good liar that Korin persuaded you he would meet Axelrod down in the registry and ask to see the evidence in the files for himself. Korin promised to show the boy he was mistaken, then bring him up to your office. Where you were waiting. Where you are waiting still, right now.'

Vasin glanced at his watch. Twenty minutes had passed since he had shot Korin. In an hour the body would start to stiffen. They had to work fast. Adamov began to answer, but Vasin spoke over him.

'I went to Axelrod's apartment, told him you wanted to speak to him. He was nervous. Axelrod and I arrived here in the basement, at the registry, as you requested. We found Korin waiting for us. Then what happened, happened. He enticed us into this laboratory. Knocked me out. Dragged Axelrod into the chamber. I recovered, and tried to switch off the machine. After a struggle, I shot him. Then I called my colleagues in the *kontora*. They informed you of the tragedy. You were the unwitting cause of Axelrod's death. But no blame will attach to you.'

The light of another *papiros* illuminated Adamov's drawn face as he dragged on it.

'No. I will not spin such lies about Korin. He did not live by lies, nor will I.'

The solidity of the story that Vasin had spun seemed to dissipate like Adamov's cigarette smoke in the vastness of the hall. He saw only Adamov's exhausted face, proud and resigned to its own destruction.

'Professor, you told me yourself. If we do not do this, you will be condemned. Removed. Your work will be undone. Petrov will have died in vain. Korin too. Without you, we are all doomed.'

Adamov sighed deeply and shook his head.

'RDS-220 is not an invention, it is a discovery. This is not a creation of any human mind, it is physical truth made real. I did not create it, I revealed it. We have discovered how to create a sun, right here on earth. It cannot be undiscovered. It will be taken to its conclusion. Not by me, but by others. I can change this one device. But I can no more stop the nuclear age than I can end fire, earthquakes, or the wind. Korin was wrong. There will always be a new Petrov, a man who sees the bomb as a path to worldly power. What they call the arms race is a race between nuclear weapons and ourselves. And they will very soon outrun us.'

A quiet followed Adamov's words. It seemed that the whole dark world inside the Professor had collapsed into a silence so deep that all future words would die in it. Vasin could think of no argument to set against the Professor's despair. To mention his own survival, Masha's, seemed trivial compared to the void that Adamov had conjured.

Masha moved toward her husband.

'Yura.' She threaded her arm into his. 'Free yourself from your cruel logic for a second. There is another logic.'

Adamov tried to push her away, but Masha only entwined herself more tightly.

'Remember what you said to me in Leningrad. When I was just a young scarecrow, and you an old goat? The goal of science is not universal truth. Instead, you said that the goal of science is the gradual removal of prejudices. A modest but relentless goal. You said that bit by bit, generation by generation, science frees men from their superstitions. And as they lose their prejudices, men see that the human world is not the center of the cosmos. Remember?'

'I remember.' Adamov's voice had softened. 'I remember, Masha.'

'You said, the discovery that the earth revolves around the sun convinced men that the earth was not the center of the universe. The discovery of microbes showed them that disease was not a punishment from God. Evolution, which showed humans that they are not some separate and unique creation of God but an animal like the rest. Remember? You said that a lot. "We are all animals, like the rest." Not gods but walking apes.'

'Apes that are bent on killing each other, child. That's the point. That is our nature. As we have discovered in this bloody century.'

'No. No, Yura. Our nature is to learn. To change. And you have spent your life creating a machine, your device, which shows men that they have finally the means in their hands to destroy themselves. Killing may be in our nature. Killing *ourselves* is not. It's the opposite of nature. Remember what Korin used to say – about that American who invented the first bomb? Oppenheimer? His new promised land where weapons would become too terrible to use? So. You reached it. You brought us all to the border of this land. After your bomb, there will be no others. But only you can bring this story to an end. Don't you see? Korin was right. Nothing can be allowed to stand in your way. Nobody. Not Petrov, not Axelrod. And Korin gave his own life for it. For you. The last blood to be spilled. Listen to Vasin. We must save ourselves. Save you. So that your precious bomb is finally tested. And then your work will be done.'

Masha ran her hand over her husband's bowed head. Adamov said nothing.

'Maybe you will come to love me as much as you loved your bombs.'

Adamov looked slowly up at his wife.

'A clever one you are, Masha. Korin always said so.'

Adamov's eyes moved from his wife's face across the carnage around him. Korin's powerful body crumpled beside the control panel. The barometric chamber with its empty window. The three spent cartridge cases glinting dully on the floor. Then he nodded, not meeting Vasin's eye, and stood.

'Perhaps I will do as you say.'

Vasin ran through the plan that he had formulated in his mind, trying to find fault with it. He knew how the *kontora* functioned: the excitement over finding an apparently real American spy with an operating secret radio would eclipse any minor inconsistencies. He faced Adamov squarely, putting all hesitation from his mind.

'Go. Now. Professor, get to your office. Don't let anyone see you going there. Take a back staircase. Call Axelrod's home number. The call will be logged. You are anxious. Don't leave until someone comes up and tells you what happened here. And when they talk to you, the men from the *kontora*, don't play dumb. You are racked by guilt that you did not see Korin for what he was. Talk to them as you talked to me. Arrogantly. You are a cloud dweller. You are a man who holds the defense of the Motherland in his hands. You are above these sordid stories. Got it?'

Adamov nodded, straightening up. He smoothed his tunic, keeping his eyes on Vasin as he worked through the story in his head, like a long equation. After a few moments, he grunted, and continued with his train of thought. Eventually he nodded once again, more to himself than to Vasin.

'A fantastical story from a paranoid mind. But it will serve. For their paranoid little minds.'

Adamov's old, imperturbable grandeur had begun to flow back into him. There was a plan to be followed. Steps to be taken. Order would be imposed once again on a world that had momentarily flown apart into a blizzard of disconnected fragments.

'Come, Maria.'

'Maria Vladimirovna will join you. You will wait together in your office. But go separately. Make sure nobody sees you on your way.'

Somewhere deep in the building a door slammed.

The three of them froze, brought abruptly back into the present danger. There was no further noise, only a suddenly oppressive sense of urgency.

'Professor, go. Masha, stay here for a moment.'

Adamov's mouth gave the faintest twitch to hear Vasin address his wife so familiarly. He looked from her to Vasin and back again, but his proud face betrayed nothing. Adamov nodded formally to both of them and stalked out of the laboratory.

VII

Maria and Vasin listened to Adamov's footsteps as they receded down the hallway. When the silence had closed about them once more, she turned to Vasin. Her face was spattered with Korin's blood. Vasin fished for a handkerchief and passed it to her. It was warm from the heat of the gun in his pocket.

Masha leaned on the console, examined her reflection in the glass of the dials, and slowly wiped off the gore.

'Better?'

'Better.'

He put out his hand and covered hers. Masha's knuckles lay under his palm like a small, trembling animal.

'You love him.'

Her face tightened into a small smile. She pulled her hand away.

'These scientists are hard to love. Every day they see perfection. And I was very imperfect.'

'Not as perfect as an equation?'

'Right.'

'Is anyone?'

She shrugged and raised her green eyes to Vasin's, steady and appraising.

'And Fyodor? He didn't compare you to the perfection of the universe?'

'Fyodor. He was a mistake.'

'Your imperfection.'

'My animal nature, Adamov would have said. But for a while I thought I loved him. Very much.'

'Adamov never knew?'

'I would have told him if I thought he would care.'

'Why wouldn't he care?'

'That was a worldly matter. And he doesn't like the world much. He loves his bombs more than any human being alive.'

'More than you?'

Masha gave a snort of impatience.

'You don't understand him. Or me. He is the greatest man I have ever met. Or you have ever met, of that I'm certain. His mind – his mind is occupied with higher things. Beautiful things. Changeless things. That's why I love him, if that's what you're asking. Adamov is a great man. He is *my* great man. You know the thieves' code. Don't be afraid. Don't ask for anything. Don't trust anyone. And don't give up your own.'

'Bombs are higher things than people?'

'Vasin, spare me your philosophizing. Adamov might need higher motives for whatever he does. Korin too. All that endless talk about the end of war, forever. Those speeches he gave you, and me. I understand, they needed that philosophy in order not to have to murder the humanity in themselves every day. To justify their work to themselves. I just want Adamov alive. And myself alive. And you gave us that, tonight. Now. You almost took it away from us when

300

you shot Korin. But then you gave it back. Nobody will ever know. But I will know.'

'Are you trying to thank me?'

'Yes. Yes, I am, Vasin. You're brave. You make your own choices. Not many men I know can say that about themselves. Not even him.' Masha gestured to Korin's corpse with a flick of her head. 'Korin was a prisoner all his adult life. The knowledge in his head. In his hands. He never had a chance to choose another path. The State would never have let him.'

'So he served.'

'He served, and he's serving still. Korin will take all the lies, all the murder onto himself from the grave. He'll even serve you.'

'Me?'

'The great investigator uncovers a spy in Arzamas. Don't say that won't bring you glory, over at your *kontora*.'

'If you think . . .'

'No. I don't think anything. You didn't bring Axelrod here for glory. You did it because Korin asked you to. Because he and Adamov took you into their confidence. They spoke to you like an intelligent man. And you chose to hear them. And believe them. And act. That's freedom, no?'

Vasin thought of the crushed heap of clothes in the chamber that had once been Axelrod, and said nothing.

'Listen to me, Vasin. All the rest . . . the glory? That's just the world. The mad world. Korin always used to say that rewards and punishment are the same. A test of vanity. Or of strength. Sent by God. Crazy old bastard. So let's say that Korin sent you a test. Luckily for you, God chose vanity.'

'If the *kontora* believes us.'

'If they believe you. You're the one who's going to be doing most of the talking, Comrade Major.'

Masha put her hand on Vasin's arm. Slight and fragile as she was, she was now the strong one. The moment when he had cradled her limp body in his arms on the roof of the Kino seemed unimaginably distant.

'It'll be okay, Vasin. I believe in you.'

Maria was about to walk out of his life. Their time was up. Mechanically, he raised his watch but could make no sense of the dial.

Vasin looked back at Masha's face. Her gaze had sharpened, and he saw that her thoughts were already striding away from him into her own private future.

'I'm glad. Glad that we are guarded by honest men.'

Masha turned and walked out of the laboratory without looking back.

VIII

In the dim silence, Vasin listened to a dull magnetic buzz that hummed through the building. The gunsmoke had dispersed, leaving a sharp smell of cordite that mingled with the hall's faint aroma of animal feces and engine oil. The pain in the back of his head, forgotten in the heat of the moment, returned with almost paralyzing force. He touched the rising swelling, sticky with blood. Good, he thought. Evidence. His blood would be on the fire axe too, and the floor. Nobody can hit himself on the back of the head.

Vasin settled himself on a stool beside Korin's cooling body and tried to concentrate on the performance that lay ahead. A landscape of deceit spooled out before him like a film that he would have to edit, carefully splicing in his fictions to arrive at this final scene of destruction. The endless patterns of intrigue joined and re-formed in his mind's eye until he lost the thread and pressed his fingers against his eyes. Absurdly, he thought of Kuznetsov, who had trusted Vasin's word that he would not discharge his weapon. Another promise broken – not that it would matter if his story held together. But would Kuznetsov himself believe his fantastical story? Of all the *kontora* men in Arzamas, it was his handler's ironic, skeptical glance that he could not quite imagine facing down as he spun his tale. But neither could Vasin imagine Kuznetsov suddenly discovering righteous indignation. He'd purse his lips, nod his beard in acknowledgment of the incomprehensible loops that life spun about him.

He thought of Masha, her physical presence, trotting up flights of stairs and peering around corners as she made her way through the deserted building to join her husband. And he thought of her words. You make your own choices, she had said. But Vasin could think of no point where he had been offered any real choice. Since he came to Arzamas he had been like a wanderer in a dream, pulling aside one curtain only to reveal another two steps behind it. And for all the secrets that he had uncovered, about Korin and Adamov in the Gulag, about Petrov and the bomb, about the forbidden loves of Masha and Axelrod, he nonetheless sensed endless acres of veiled, forbidden knowledge still surrounding him, stretching into darkness.

Vasin felt loneliness seating itself beside him like a companion who doesn't need to speak. He would never see Masha again. The only person who had ever called him brave, or probably ever would. What had he wanted from her? To make her his mistress? To escape with her into a different future from the one that the world had prescribed for both of them? Now that she was gone, Vasin realized with a sharp pang that, yes, that was the secret his own heart had kept veiled, even from himself.

There had been a clarity to Masha, an animal single-mindedness that Vasin found obscurely shaming. Her childhood suffering, the violence that she had inflicted on others to survive, the ruthlessness with which she had used and deceived him in defense of her Adamov. Even the addled moment when she decided to destroy herself: all these were impulses of absolute, fearless resolution. Masha had been a mirror upon which Vasin's life had been violently dissected. And he could find no absolutes to put opposite her own. His own life had been a series of useless efforts, each driven by the material dictates of the little world that surrounded him, or the pathetic impulses of his body. But Masha was the one who truly did not live by lies.

In a few minutes, from the moment that he picked up the

telephone and invited the world to burst in on his silence, Vasin would be irretrievably plunged back into the tangled unhappiness of his own life. And he realized that perhaps unhappiness was the one state he truly deserved. He was not a cloud dweller, but a swamp dweller. There were no pure universes of numbers in his life, no eternal truths to discover. Just existence, with its daily compromises. But at least, here in Masha's death-dealing city, he had touched a different world. A world from which Vasin had brought away a lie of his own, the falsehood of Korin's espionage, to add to all the rest of the lies in the basements of the *kontora*. But this would be *his* lie. A good lie. And Vasin would share this secret with Masha, and with Adamov, and it would bind them together forever. A secret that he would know, and Masha would know, but that the *kontora* would never know. Which felt almost like a victory. And that gave him strength.

In front of Vasin was a telephone, its wire linked to other wires that spread across the secret city and out across the great Soviet empire like an infinite web. Vasin waited for another minute, feeling time run around him like a stream.

Then he picked up the receiver, and dialed.

CHAPTER ELEVEN

MONDAY, 30 OCTOBER 1961

THE DAY OF THE TEST

Vasin glanced at the clock on Zaitsev's wall.

10:35.

An hour to the test.

Up in Olenya, snow would be swirling in the monstrous draft of the Tupolev bomber's propellers as the pilots prepared to taxi for takeoff. The military's top brass would be there, shivering in the Arctic wind. Korin's loading team would be huddled in the lee of their fuel truck, watching the plane's run with hard, unimpressed stares. The Sailor might even be among them, chewing on an unlit *papiros* cigarette in the slanting morning light.

There was a gentle knock on the door of Zaitsev's office. Hesitantly, Efremov entered, bearing an armful of dossiers. The adjutant's former cold hauteur had dissolved into nervous hesitation.

'Major? The documents you requested.'

Vasin gestured to the sea of paper that already covered the General's conference table. Reverently, as though they were sacred objects, Efremov placed the files with the rest.

'Vasin. I just wanted to say . . .'

'Make it brief, Efremov.'

Vasin was now the spy catcher. The bloodied executioner. He was now a man with no time to spare for the likes of Efremov.

'I wanted to congratulate you, Comrade. Wanted to say that I was always on your side. You should know that it was Zaitsev who insisted on placing obstacles . . .'

'Anything else?'

Efremov's angular face had gone pale. He drew himself up and saluted. Vasin returned the salute with a casual flick of the hand.

'One thing before you go. Kuznetsov was an excellent choice as my handler. Helped me a lot. I'm recommending him to the higher-ups for promotion. A posting to fraternal Cuba, we're thinking. I knew you'd want to congratulate him before he goes.'

General Zaitsev himself was off supervising the search of Korin's barrack. The last twenty-four hours had drained him of his habitual choler like a bloodletting, leaving only pale nervousness behind. In the cold light of the previous dawn, as Zaitsev and Vasin had faced each other on the steps of the Institute, the old brute had looked deflated. A spy. Oh yes. A real American spy in the heart of Arzamas. And Zaitsev had failed to discover him. The knowledge of his impending disgrace had punctured the General like a balloon. Zaitsev's enormous uniform seemed to hang on him like a sack. And in his eyes, when they met Vasin's, was pure, animal fear.

Vasin had called Orlov at home from the secure line in Zaitsev's office, which by the unspoken right of victory had temporarily become his own. It had been half past five in the morning, but Orlov was already awake. Or perhaps still awake. Vasin had communicated only the essentials. Colonel Korin, a spy and double murderer. Religious fanatic. Secret radio. Shot dead. Requesting instructions.

'Understood' was all Orlov said. The silence that followed lasted a minute. 'A team will be at the Arzamas airfield in four hours. The witnesses are to speak only to Special Cases. Stand by.' Then the electronic purr of the disconnected line.

Zaitsev's clock ticked forward. 10:55.

The Tupolev bomber would be climbing steadily toward the testing ground now, laboriously gaining altitude. Adamov was in the radio room at the Citadel, listening in to the bombardiers' and pilot's reports. He'd been there since dawn. A pair of Vasin's Special Cases comrades were discreetly escorting the Professor wherever he went to ensure that none of Zaitsev's goons tried to speak to him.

Vasin wondered how Adamov had pretended to take the news of Axelrod's death, of Korin's, communicated by some stammering *kontora* minion. With superb unconcern, he would guess. He could imagine Adamov's slow blink, the magisterial nod that acknowledged the latest sordid affair of the world. Vasin had little doubt that the Professor would play his role perfectly.

At Arzamas's airfield, a *kontora* plane was waiting for them. Orlov's terse orders: Fly to Moscow immediately after the test. Bring Adamov, but do not speak to him beyond pleasantries. Gather the most important files on the Petrov murder and take them with you. Seal the rest. Brief the Special Cases counterintelligence team who will remain in Arzamas.

Crystal clear. Orlov's order, imposed on chaos.

But first, the test.

11:22.

Zaitsev's secretary came in with tea, which she placed on the table in front of Vasin with exaggerated formality before backing away. Vasin did not acknowledge her. He was staring out of the window over the rooftops of Arzamas. A bright autumn sun had burned off the morning's mist, leaving the sky a deep blue with a marbling of cloud. Somewhere far to the north, beyond the curve of the earth, the bomber crew would be arming RDS-220 for detonation. The pilots would be making their final reports to ground control as they prepared for their approach to the test site.

11:31.

Between Vasin and the bomb were thousands of kilometers of clear, bright air, a universe of trillions of invisible molecules, all vibrating to a mysterious, unheard rhythm. He thought of RDS-220 tumbling from the sky, momentarily free of its cellars and its bindings, falling beautifully through the morning sky, accelerating downward as the earth pulled it toward herself.

The minutes ticked by. Starlings wheeled around the domes of the old monastery. Foolishly, Vasin found himself straining his ears to listen. His fingers closed on the edge of the desk, bracing. But the air did not burst into flame. The hand of the clock moved slowly on. And the earth continued to turn, moving Arzamas slowly toward the noonday sun. Somewhere, beyond the horizon, Adamov's black sun ignited its own terrible dawn.

CHAPTER TWELVE

TUESDAY, 31 OCTOBER 1961
THE DAY AFTER THE TEST

'My dear fellow!'

Orlov sprang from his desk and took Vasin's hand in both of his. The General's face was animated with a grin of triumph. Holding Vasin by the arms, he looked his protégé up and down, as though checking that Arzamas had returned him in one piece. Orlov turned Vasin half around and inspected the thick dressing on the back of his neck. Vasin met his chief's eye, searching for any dancing spark of anger that would betray that word of his affair with Katya had reached Orlov. But he saw nothing other than a glow of pride in the General's face.

'Our wounded hero! Scoundrel nearly knocked your head off its neck, I'm told. But my boys are tough. Tough as nails.'

Orlov squeezed the bandage hard, bringing tears of pain to Vasin's eyes.

'Sit! Sit.'

The General steered Vasin into a chair, then bounced down into his own.

'A remarkable triumph. And yet you said nothing, all these days. Nothing about your suspicions. Quite the dark horse you are, Vasin.'

Orlov's chestnut eyes scrutinized Vasin's face with the intensity of a searchlight.

'Didn't wish to raise any false accusations until I had evidence, Comrade General.'

'Naturally.'

'Given the sensitivity of the charges. Sir. And the positions of the suspects.'

'Of course. You acted correctly.'

Orlov's rare smile remained switched on, unwavering as a light-bulb. He waited for Vasin to continue.

Vasin smiled back, with suitable modesty, but said nothing.

'Pavel Korin,' Orlov continued eventually. 'Who would have thought? I read his file, of course, as soon as his name came up in the Petrov investigation. Some doubtful episodes in Korin's past, of course. But a spy? Well. That came as a surprise. Of course all the clues to his treachery are there, if you look for them with the right eyes. A bacillus, introduced by our American so-called allies, let loose in the very heart of our defenses at the very moment that we were supposedly fighting side by side. Yes. Your story tracks well. I cannot fault your scenario, Vasin.'

'My scenario, sir?'

Vasin felt his mouth go dry. Had Orlov guessed the truth? If so, the General's poker face gave nothing away.

'Your investigator's logic, I mean.'

'Indeed, sir.'

'I will assign an operative group to investigate the damage Korin may have done over his career of treachery. I have no doubt that they will come up with much that is useful to me. And we will be reviewing your debrief in detail over the next few days. If your doctors permit it.'

'Even if they do not, sir, I am ready.'

'Good man.'

'And Adamov, sir? You have spoken to him?'

Adamov. During the flight from Arzamas to Moscow, Vasin had not spoken a word to the Professor. But they had exchanged a long look. Of complicity? Thanks? Resentment? Vasin had no idea what Adamov hid behind his grave, fierce stare. Up there, among a bright tumult of clouds, Adamov was in his natural element. Once more they had become men of different worlds.

Orlov's smile did not flicker, though he did not answer immediately.

'The Comrade Professor is being most cooperative. Though of course he is also busy receiving the congratulations of Comrade Khrushchev and his colleagues at the Academy. For his brilliant work.'

Vasin had heard the official announcement on a radio in the KGB sanatorium that morning. The Motherland's new bomb, a terror to our enemies, a shield that will protect our socialist home from aggression. A bomb to strike fear into the capitalist cowards.

'I am glad. I was afraid he would have mixed feelings. Korin was the Professor's old friend.'

'Korin was a friend to many, Vasin. To many. He was a deceiver. Ruthless. Clever. A most dangerous enemy.'

'And the Professor's wife? She is well?'

Something sly and pointed had crept into Orlov's smile.

'Interesting that you should ask. I believe she is well. She assisted you during your investigations?'

Vasin shifted uncomfortably on his chair but did not answer.

'Sir, may I ask you a question?'

'You may.'

'Why did you send me to Arzamas? What did you think I would find there?'

'Ah, Vasin. You flatter me. You think I know everything in advance.'

'But you had something in mind? Somebody?'

Orlov gave an exaggerated shrug.

'You have earned the right to know my thoughts. So I share them with you. Perhaps you will learn something from them. It is very simple. Petrov was a golden child. The son of a man who has every chance of becoming the President of the Academy of Sciences. The young man kills himself. Perhaps. But, why? A girl is not interesting to us. Depression? Likewise. However, maybe there is something more to it.

Something that his father would prefer to keep hidden. So. As guardians of so many uncomfortable secrets, we in this office have the duty to discover what happened. Strictly in the interests of State security, of course. What if someone else, some enemy, discovered a sordid secret about Fyodor Petrov? This would give them power over his father, one of our most respected scientists. We will not allow this to happen. This much you guessed already, I suppose?'

Vasin nodded obediently.

'And if it was not suicide but murder? Well, even more interesting. The murder weapon was so exotic. Almost the bite of a speckled band snake. I see from your smile that you know your Conan Doyle. Good. So, who would use this rare, radioactive poison? Only a colleague. Obviously. Perhaps a powerful colleague. Someone in authority who no longer deserves the trust of the Motherland. And if you wish to ask, did I suspect Adamov, my frank answer is no. Not specifically. I had no knowledge of his personal history with Petrov's father. But did I think that such a story could be behind this affair? I did.'

'So you sent me on a fishing expedition?'

Orlov's unnatural bonhomie finally evaporated, replaced by his usual scowl.

'Naturally. That is what I do, Vasin. I fish. Sometimes with trawls. Sometimes with flies. Sometimes with baited fish traps. And you are my obedient little fly. My very obedient fly.' The General's gaze wandered to the glass paperweight with its eternally trapped dragonfly that sat on his desk.

Suddenly, out from under their deeply hooded lids, Orlov's eyes flicked up to meet Vasin's.

'Only the weak hate. You know that, don't you, Vasin?'

'Sir?'

'The weak hate. The stupid hate. The strong act. The clever act, but not always immediately. The intelligent keep score. They keep accounts.'

An unmistakable note of menace had crept into Orlov's voice.

'Not sure I follow you, sir.'

'Would you like to know where my dear Katya's last two lovers are now?'

Vasin froze. Even the throbbing of his injuries disappeared in the suddenness of his shock. Orlov straightened in his chair, not releasing Vasin from his angry stare.

'You would like to know, wouldn't you? How rich is your imagination, Vasin? Tell me. *Tell me.*'

Orlov's voice had sunk to a low hiss, and his eyes glistened with a sadist's glee.

'Nothing to say for yourself? You disappoint me.'

The General leaned forward and glanced frankly down at Vasin's crotch.

'The last one pissed himself. Right here. Lost control of his bladder. Imagine! What mere words can do to a man. But then, you knew that. At least at second hand. You've read some of the files. You know what we are. What we do.'

Vasin felt the office swim before his eyes. He clutched the arms of his chair for support, felt the polished wood digging into his palms. He felt the vertigo of a man standing up against an execution wall. Counting bricks. Counting breaths.

'Sir. She . . . I . . .'

'Did you like it, Vasin?' Orlov's voice had dropped to a soft, hissing whisper that was almost sexual. 'Did she moan like a whore? Go on. Say it. Did you think of me when you fucked her?'

Vasin's eyes were pleading. Could he attempt to apologize? Tell Orlov that he had been seduced? Or was it time, finally, to let his anger and humiliation explode? To scream and rage at the evil and injustice of this accursed place? Of this man?

Orlov's breathing had become shallow. A flush of color had come into the General's smooth, priestly face. The intensity of his stare had become almost carnal as he watched Vasin twist and wilt under his power.

'Tell me what you think of me, Vasin. Say it.'

Vasin fought back words as though choking back vomit. You cynical monster. You sadist.

'You are a strong man. A wise man.'

'Good. Very good. And my wife, Katya? Who is she?'

'She is a shameless whore, sir.'

'Yes.'

'And us. You and me. Who are you to me, now, if I choose to forgive you?'

'Your loyal servant, sir.'

'My loyal servant?'

'Your very obedient fly.'

'Again.'

'Your very obedient fly. Sir.'

'Excellent.'

Orlov sighed deeply and subsided back into his chair.

'Oh, Vasin. Oh, my boy. I was so hoping that you would see wisdom.'

'Wisdom?'

'That spirit in you. Intelligence. Independence. Katya saw it too. She said, That Vasin, he's a smart one. Make him one of yours. She has a good eye.'

Vasin felt his mouth go slack.

'Yes. You are surprised? You think that any wife of mine would dare to defy me? Could get away with deceiving me? How sordid that would be. How pathetic. No, Vasin. She is one of mine. She tests. She tastes. She whispers in your ear, What do you really think of my husband? Isn't he a pig? A fool?'

Katya's bedroom words, exactly. Vasin winced at the memory.

'You know Katya was once a rebel, too, just like you. Back when she was wild and pretty. But you see, I need rebels, Vasin. Men who can think for themselves. Minds who can see beyond the system. But not rebels who rebel against *me*. You see that, don't you?'

The pain in Vasin's neck returned as a pulse of agony. It was almost as though someone was sliding a great hook into his flesh.

'Yes, sir.'

'I think we are ready, don't you?'

'Ready?'

'For the next level, Vasin. Your next assignment. But you must be very secret.'

Orlov paused to savor the moment. He pulled a thin file marked TOP SECRET from a drawer of his desk and passed it to Vasin.

'Yes. I have some even more surprising news for you, Comrade. Or as I will soon be calling you, Lieutenant Colonel Vasin. A great task awaits you. You see, your revelations about Colonel Korin could complete a puzzle on which we have been working for some time.'

'Sir?'

'A lead we have been working on for months. It seems we have a traitor in our midst. A spy in the very heart of State Security. Yes, Vasin. He is *one of us*. But this man has a powerful protector. There is no direct evidence against him. But your spy Korin may be exactly what I need to collect that evidence. Especially since Korin is conveniently dead.'

'Conveniently?'

Vasin's voice had become a whisper.

'Dead men tell whatever tale the living place in their lifeless mouths, Vasin. As I suspect you know already. Are you ready to find me some more tales for Korin to tell from beyond the grave? Specifically – the identity of the traitor Korin's controller?'

Live not by lies. Vasin's own words sounded in his mind like a mocking echo.

'I am ready, sir.'

'Good. Very good. Now go. Your family is waiting for you. Vera will have missed you.'

Vasin struggled to his feet. He found his head bowing down with the weight of its burden of deception. Orlov stood also and surveyed his new creature with satisfaction.

'Welcome home, Colonel Vasin.'

A NOTE FROM THE AUTHOR

Black Sun is based on a true story.

At 10:50 Moscow time on the morning of 30 October 1961, a specially adapted Tupolev-95 bomber took off from Olenya air base carrying the most powerful weapon ever created by mankind.

The twenty-seven-ton, twenty-six-foot-long device's code name was RDS-220. American journalists later nicknamed it the Tsar Bomb. But RDS-220's actual creators took it far too seriously to call the bomb by anything but its real name. In truth, they feared it. In the weeks leading up to the test-firing of RDS-220, the bomb's real-life designer, academician Andrei Sakharov, of whom my fictional Yury Adamov is a dark twin, became concerned that his new device was so powerful that it might cause a runaway chain reaction in atmospheric hydrogen, or possibly nitrogen. Sakharov ordered a team of his engineers at Arzamas-16 to calculate the chances that the detonation might actually set the earth's atmosphere on fire.

Earlier in his career, Sakharov had speculated about the theoretical possibility of even larger bombs, of two hundred and five hundred megatons. But he was so shocked by the results of his colleagues' theoretical conclusions about the unpredictable effects of RDS-220 that he made a radical decision. Ten days before the test, he ordered the device's revolutionary new uranium tamper to be replaced with a lead one. The bomb makers of Arzamas-16 had finally touched the outer limit of science. They had created a bomb too powerful for the earth to withstand. And they stepped back.

Even with a specially extended runway, the Tupolev took off with difficulty. The bomb's weight was twice that of the aircraft's usual payload. Both the release plane and a Tu-16 observer aircraft that was to take air samples and film the test had been painted with special reflective white paint to minimize heat damage. But despite this precaution, the chances that the crew would survive the test were put at 50 percent. The Tupolev's pilot, Major Andrei Durnovtsev, had been informed of the risk. The rest of his airmen had not.

Shortly after 11:30, Durnovtsev had reached Mityushikha Bay, a nuclear testing range in the Novaya Zemlya archipelago in the Arctic Ocean. The film of the test shows a desolate landscape of snow and rock. At 11:32 Moscow time, at an altitude of 10,500 meters above Zone C of the Sukhoy Nos section of the test site, he released the bomb. Its fall was slowed by specially made parachutes designed to allow the plane time to get a safe distance from the detonation.

The altimeter-activated firing mechanism of RDS-220 detonated perfectly at four thousand meters. The fireball nearly reached the altitude of the release plane and caused both aircraft to tumble more than a kilometer. In the official film, the cameraman in the Tu-16 observer plane struggles to maintain focus on the detonation during the free fall. But both aircrews survived.

The explosion destroyed every building, both wooden and brick, in the evacuated village of Severny, some 55 kilometers from ground zero. The heat from the blast, according to unmanned sensors, was still strong enough to have caused third-degree burns 100 kilometers away. The thermal pulse was felt by human observers 270 kilometers distant. The shock wave broke windows in Norway and Finland, some 900 kilometers from the test site.

The mushroom cloud eventually rose to an altitude of 64,000 meters, over seven times the height of Mount Everest and far above the earth's stratosphere. The cap of the mushroom cloud had a peak width of 95 kilometers and its base was 40 kilometers wide. It could be seen over a thousand kilometers away. Earthquake sensors in

America and Japan registered the blast as a seismic wave measuring 5.5 on the Richter scale, and the shock waves were still measurable on their third circuit of the earth.

Sakharov and his team later estimated the yield of RDS-220 at 56 million tons of TNT, the equivalent of 3,800 Hiroshima bombs detonated simultaneously, or ten times the combined energy of all the conventional explosives used in World War II. The zone of 'total destruction', formally defined by Soviet military planners of the day as the complete annihilation of all buildings and life, was 70 kilometers across, roughly equivalent to the entire area of metropolitan Paris.

The world was shocked. Sakharov was shocked. The power of RDS-220 – even in its reduced form – was so enormous that Sakharov, the father of the Soviet hydrogen bomb, refused to ever work on another device. He began to campaign, first within the Academy of Sciences and then publicly, for a total ban on atmospheric nuclear testing. It was to be the first of many, increasingly existential, battles that Sakharov was to fight against Soviet power. Amazingly, he was ultimately to win them all, though at the cost of his career, his privileges, and his health.

Sakharov's first victory came in 1963, when both the United States and the USSR agreed to sign the first Nuclear Test Ban Treaty. The world's nuclear powers continued to test much smaller nuclear bombs underground and under the sea, but never again was a nuclear device detonated in the atmosphere. Sakharov went on to become the Soviet Union's most prominent dissident, applying his fearless physicist's rigor to his attacks on the amorality of the Soviet State. He successfully campaigned for the right of dissidents and Jews, the so-called refuseniks, to leave the USSR. During Mikhail Gorbachev's era of glasnost, Sakharov became a passionate and authoritative voice calling for a frank appraisal of the Soviet crimes of the past. Sakharov was awarded the Nobel Prize for Peace in 1975. In 1991 the Soviet Union collapsed.

My fictional Adamov is not Sakharov, though both Adamov and Korin give voice to the paradoxical hope argued in Sakharov's

Memoirs, that they are building bombs in order to create world peace. Adamov's story is borrowed from the lives of two giants of Soviet science: Nobel Prize winner Lev Landau, who pioneered much of the quantum physics used in the hydrogen bomb, and Academician Sergei Korolev, the father of the Soviet manned space program.

Landau was born to a Jewish family in Baku in 1908. He studied with Niels Bohr, one of the fathers of nuclear physics, in Copenhagen before voluntarily returning to Stalin's Russia in 1932 to head the Department of Theoretical Physics at the Institute of Physics and Technology in Kharkov, Ukraine. Under Landau, the Kharkov Institute became one of the world's leading centers for nuclear physics; the discoveries listed in my fiction under Adamov's name in reality belong to Landau. Like Sakharov, and like Adamov, Landau was fearless in his moral judgments of the Soviet State, despite working in the heart of its defense establishment. In 1956 the KGB noted that Landau described the Soviet crushing of the Hungarian uprising as 'red fascism'. But like all the cloud dwellers of the Soviet postwar nuclear program, Landau was allowed to voice his views freely, as long as they remained inside the secret world that he inhabited.

Korolev's career as the father of Soviet rocketry was no less stellar but was almost destroyed by Stalin's paranoia. Like tens of thousands of Soviet scientists, Korolev found himself denounced and sent to the Gulag in 1938, during the hysteria of the Purges, on spurious charges of sabotage. He was eventually transferred to a *sharashka*, a prison research laboratory, in Moscow, but not before nearly dying of scurvy and losing all his teeth in a gold mine in Kolyma. His brilliant colleagues at the Jet Propulsion Research Institute were less lucky: Both his bosses were executed. In the *sharashka* known as Central Design Bureau 29, Korolev worked alongside the famed Soviet aircraft designer Andrei Tupolev. From their prison laboratory they created the Tupolev Tu-2 bomber and the Petlyakov Pe-2 dive-bomber, as well as rocket-assisted takeoff boosters for aircraft motors. Korolev was finally allowed to return to his family in 1944, though friends noted that his character was profoundly damaged by the experience, and charges against him were not dropped until 1957. Korolev went on to create *Sputnik 1*, the world's first orbiting satellite, and to put Yury Gagarin into space in April 1961, a few months before the test of RDS-220. My fictional Adamov and Korin stand for a whole generation of leading Soviet scientists and engineers who spent some of their best years in prison.

Maria Adamova is also based on two extraordinary Soviet women. One of them is Ekaterina Segeyevna Gvozdeva, whom I met in Leningrad in August 1991 during the tumultuous days of the failed hard-line coup attempt that was to bring a final end to Soviet power. Gvozdeva, known as Kitty, was born in St Petersburg in 1899. She married the rector of Petrograd University and in the 1920s played hostess to some of the city's leading intellectual lights, including the poets Anna Akhmatova and Alexander Blok. She remained in her native city throughout the nine-hundred-day Nazi siege, enduring the horrors of entire families of her neighbors dying of hunger. The story of Masha's murder of the man who tried to rape her actually happened to Kitty's best friend, and was recounted, matter-of-factly, by Kitty as she cut into rock-hard pears she had bought at the market with half of her monthly pension.

The other woman on whom Masha is modeled is my mother. Born into the family of a senior Party apparatchik in Kharkov, Ukraine, in 1934, my mother and her sister, Lenina (named for Lenin), had their privileged world turned upside down by the arrest of their parents during the great Purge of the Party in 1937. Her father, unbeknown to her, was almost immediately shot on spurious charges of sabotage; her mother was sent to a prison camp in Kazakhstan for fifteen years; there she went mad. My mother and her sister were first sent to a children's prison and then to an orphanage to be reeducated as model Soviet citizens.

During the German advance across Ukraine in the autumn of 1941, Lenina was mobilized with the older children to dig trenches. My mother, aged seven, was put on a raft with the younger orphans and set adrift on the Dnieper River to float to safety. For much of the following year my mother was one of the thousands of wild, starving, homeless children carried eastward on the winds of war. She spent the bitter winter of 1942–43 on the east bank of the Volga near Stalingrad, gathering cigarette butts to sell to soldiers for food. She was eventually evacuated to a vast orphanage in Solikamsk in the Urals, where by miraculous coincidence she was reunited with her sister.

My mother remembers little of that time, except to say that she lived by 'wolves' laws'. And like Maria Adamova, she found her salvation in study, winning a place at Moscow University and graduating at the top of her class. She later met a British academic and, after a six-year struggle against the Soviet authorities, married him. The world my mother made for her adult self was one of books, intellectual friends, a passion for ballet and beautiful things. But the memory of childhood hunger and violence never left her. But unlike Masha's, my mother's spirit never wavered.

Owen Matthews
Wytham Abbey, Oxfordshire, November 2018

ACKNOWLEDGMENTS

Many people helped this book in its journey onto the page – first and foremost my dedicated and patient editor, Robert Bloom at Doubleday, who worked tirelessly above and beyond the call of duty to shape and polish *Black Sun*. Without my genius agent Toby Mundy I would never have hooked up with Rob, or thought to make Alexander Vasin the hero of a trilogy. And without the encouragement of my friends I would never have imagined writing my first thriller in the first place. Particular thanks to Lisa Hilton, Jonny Dymond, Andrew Jeffreys, Charles Cumming, and Richard Stow for all their help and support while the book was being written. And to Xenia, Nikita, and Teddy for putting up with a writer in the house, which is no easy fate. Finally, a bow to the real Alexander Vasin, my late uncle, who as a young Soviet tank commander lost a leg outside Smolensk in 1944 but despite his disability rose to be the USSR's deputy minister of justice and a wise and loving husband to my aunt Lenina.

ABOUT THE AUTHOR

Owen Matthews reported on conflicts in Bosnia, Lebanon, Afghanistan, Chechnya, Iraq and Ukraine and was *Newsweek* magazine's bureau chief in Moscow from 2006 to 2016.

He is the author of several non-fiction books including *Stalin's Children, Glorious Misadventures* and *An Impeccable Spy*. He lives in Moscow and Oxford.